PROBLEMS OF REPRESENTATION IN THE TEACHING AND LEARNING OF MATHEMATICS

PROBLEMS OF REPRESENTATION IN THE TEACHING AND LEARNING OF MATHEMATICS

edited by

Claude Janvier

University of Québec at Montréal

LAWRENCE ERLBAUM ASSOCIATES, PUBLISHERS

1987 Hillsdale, New Jersey London

Lawrence Erlbaum Associates, Inc., Publishers
365 Broadway
Hillsdale, New Jersey 07642

Library of Congress Cataloging-in-Publication Data

Problems of representation in teaching and learning math.

Papers derived from a symposium organized by CIRADE of Université du Québec à Montréal.
Bibliography: p.
Includes index.
1. Mathematics—Study and teaching—Congresses.
I. Janvier, Claude. II. Université du Québec à Montréal. Centre interdisciplinaire de recherche sur l'apprentissage et le développement en éducation. III. Title: Representation in teaching and learning math.
QA11.A1P75 1986 370.15'6 86-6234
ISBN 0-89859-802-8

Printed in the United States of America
10 9 8 7 6 5 4 3 2 1

Contents

Introduction

As the title suggests, this book presents more questions than answers. Although it was prepared for those involved in research in the field of mathematics education, we also had in mind our colleagues who assume responsibility for the training of teachers. It is aimed at providing those concerned with basic research orientations, a fund of fundamental theoretical perspectives, research methodologies, analyses and research results.

This book stems from a symposium organized by CIRADE (Centre Interdisciplinaire de Recherche sur l'Apprentissage et le Développement en Education) of Université du Québec à Montréal. Because the structure of the book and its contents derive from this symposium, I describe it briefly.

One of the research groups in the CIRADE was approaching the final stages of a research project on representation. This group thought it would be helpful to invite colleagues for mutual discussion in an attempt to develop a more comprehensive theoretical framework. The participants were invited on the basis of their involvement in the domain and on their capacity to share their past reflections. The participants were not invited to state what they already knew, but to identify through discussions what they still did not know. The discussions were initiated by the reactions of the participants to papers circulated prior to the conference.

Those papers, chosen on the basis of the coherence and scope they collectively provide, make the first part of this book. They can be regarded as presenting a "state of the art" of representation in mathematics when the conference met in 1984.

This collection of articles, after two basic introductory chapters, can be divided in two sets. The first deals with a central issue, namely, the notion of translation. Each chapter of the second set addresses differently the delicate question of internal representations.

The first introductory chapter is from Ernst von Glasersfeld. It was selected to set the scene with a constructivist theory of knowledge (departing from behaviorism) that is generally shared by all researchers in the domain of representation. This explains why the chapter is longer than the others, although it could be considered short in view of all the variables it has to take into account. In fact, the reader does not find any mention of issues related to the use of representations. Von Glasersfeld's chapter rather paves the way to the understanding of all the others. It presents a deep analysis of the fundamental concepts that are omnipresent in the book and that require re-examination. As he puts it: "For about half a century behaviorists have worked hard to do away with mentalistic notions such as: meaning, representation, thought." Clearly, a genuine concern for the notion of representation implies, firstly, a renewed analysis of stimuli and responses (stressing the importance of the configurations involved) and secondly the abandonment of the black box approach to the study of mental processes. Von Glasersfeld's theoretical position singles out how the prevailing "communication of knowledge" paradigm is misleading when teaching and learning processes are to be examined. Indeed, von Glasersfeld points out the active role of the learner and his (her) affective involvement. Towards the end, he mentions that one learns through failure or surprise. But, as he says, failure or surprise require some expectation: "From the constructivist point of view, it makes no sense to assume that any powerful cognitive satisfaction springs from simply being told that one has done something right, as long as rightness is assessed by someone else. To become a source of real satisfaction, rightness must be seen as the fit with an order one has established oneself."

Afterwards comes Jim Kaput's chapter in which we find most of the questions relevant to the domain of representation in mathematics reasoning. It can be considered as the state of the art in 1983. Most (if not all) fundamental issues regarding the use of external representations (symbolic configuration) are touched upon. The need for an adequate theory is very well substantiated. We notice that the first outline developed is at times at variance with von Glasersfeld's chapter. The conference pointed out clearly the need for a better integration of the von Glasersfeld's and Kaput's positions. In the last section, he examines seven situations in mathematics proper in order to illustrate the representational nature of mathematics itself.

The next set of chapters addresses the particular question of translation. Other chapters tackle the issue, namely Kaput's and Mason's. However, those three chapters consider solely this question. It should be mentioned that the chapter of Lesh et al. (Representations . . . , 1985) is acutally a revised version of a previous paper but does not include the conference concerns. Together with Janvier's chapter it contains theoretical considerations based on previous research on rational numbers (Lesh et al.) and on function (Janvier). They actually tackle the question almost "in parallel." They provide each a diagram, comment on the role of (spoken) language as facilitator, mention the translation within the same mode of representation, mention their unstable character . . . They lead both to chapter five by Lesh et al. providing interesting research results.

The following three chapters tackle the difficult question of internal representations. Gerald Goldin, in the analysis of mathematical problem solving, considers fundamental what he describes as imagistic processing. As he says: "imagistic . . . is not intended to suggest purely visual images, but to have the connotations of the word "imagine." Claude Janvier introduces the idea of semantic domain. He suggests that a variable and a transformation are two different internal representations for the only mathematical concept of function. He makes an attempt at defining representations in order to incorporate internal representation. As for John Mason, his chapter is based on five examples. The "palpability of symbols" is studied using notions such as "subvocal inner speech," "acoustic responses," "pictural and kinesthetic images." The voyage metaphor is used to illustrate the personal construction of meaning and the role of imagery.

With diSessa's chapter we make the first real plunge into the realm of internal representation. The article was selected on the basis that it points towards new research directions. In fact, it was hoped that the general theoretical perspective used by diSessa could be applied to mathematics reasoning.

Indeed, not only could phenomenological primitives exist in mathematics but they could as well share properties with those investigated by diSessa.

The second part presents a selection of excerpts from the symposium. They are reactions to and reflections on the chapters appearing in Part I, which we assume have been read. The choice was made so that each excerpt is relatively self-contained. Chapter 11 highlights certain questions that were partially discussed during the conference. As the three colleagues—Bernadette Janvier, Nadine Bednarz, and Maurice Bélanger of the CIRADE—say at the beginning of their chapter, they would like to "contribute an analysis of some of (their) past and current research on the use that is made of representations in the teaching of mathematics and their effects."

Part III contains chapters that were sent later by the participants. As expected, the conference brought all participants further in their reflections. The strategy of "forcing the dialogue" was successful and has produced results. Indeed, the reader finds that some obscure issues became clarified or were set into a new and more illuminating theoretical perspective.

Several questions received answers but a lot more stemmed from the confrontation of positions. We hope that this feature contributes to making the book more relevant for all those involved in research.

ACKNOWLEDGMENTS

I make grateful acknowledgment to the Université du Québec à Montréal and to Québec's Ministry of Education for having sponsored the Conference. I also want to thank those who helped me in improving the English: Maurice Bélanger, Julia Hough, and Jim Kaput. Special thanks go to Louise Poirier Saintonge who worked hard to make the transcripts, and to Sophie René de Cotret who helped me with the subject index. My last thought goes to all the researchers (colleagues, assistants) of the CIRADE who made the Conference a success.

I FIRST STEP INTO THE PROBLEMS

1 Learning as a Constructive Activity

Ernst von Glasersfeld
University of Georgia

Ten or 15 years ago, it would have been all but inconceivable to subject educators or educational researchers to a talk that purported to deal with a theory of knowledge. Educators were concerned with getting knowledge into the heads of their students, and educational researchers were concerned with finding better ways of doing it. There was, then, little if any uncertainty as to what the knowledge was that students should acquire, and there was no doubt at all that, in one way or another, knowledge could be transferred from a teacher to a student. The only question was, which might be the best way to implement that transfer—and educational researchers with their criterion-referenced tests and their sophisticated statistical methods were going to provide the definitive answer.

Something, apparently, went wrong. Things did not work out as expected. Now there is disappointment, and this disappointment—I want to emphasize this—is not restricted to mathematics education but has come to involve teaching and the didactic methods in virtually all disciplines. To my knowledge, there is only one exception that forms a remarkable contrast: the teaching of physical and, especially, athletic skills. There is no cause for disappointment in that area. In those same 10 or 15 years in which the teaching of intellectual matters has somehow foundered, the teaching of skills such as tennis and skiing, pole jumping and javelin throwing has advanced quite literally by leaps and bounds. The contrast is not only spectacular but it is also revealing. I return to this phenomenon at a later point when, I hope, we are able to consider an analogy that, at this moment, might seem utterly absurd.

If educational efforts are, indeed, failing, the presuppositions on which, implicitly or explicitly, these efforts have been founded must be questioned, and it seems eminently reasonable to suggest, as did those who formulated the topic for this discussion, that we begin by inspecting the commodity that education claims to deal in, and that is "knowledge."

This chapter is an attempt to do three things. First, I consider the origin of the troubles we have had with the traditional conception of knowledge.

Second, I propose a conceptualization of "knowledge" that does not run into the same problem and that, moreover, provides another benefit in that it throws helpful light on the process of communication. In my experience, this is an area that has not been given much thought. Educators have spent and are rightly spending much time and effort on curriculum; that is, they do their best to work out what to teach and the sequence in which it should be taught. The underlying process of linguistic communication, however, the process on which their teaching relies, is usually simply taken for granted. There has been a naive confidence in language and its efficacy. Although it does not take a good teacher very long to discover that saying things is not enough to "get them across," there is little if any theoretical insight into why linguistic communication does not do all it is supposed to do. The theory of knowledge that I am proposing, though it certainly does not solve all problems, makes this particular problem very clear.

Lastly, having provided what I call a *model* of "knowing" that incorporates a specific view of the process of imparting knowledge, I briefly explore a way to apply that model to the one thing all of us here are interested in: how to introduce children to the art, the mystery, and the marvelous satisfaction of numerical operations.

THE INSTRUMENTALIST ANSWER TO THE SKEPTIC'S ATTACK

The nature of knowledge was a hotly debated problem as far back as the 6th century B.C. The debate has been more or less continuous, and although in many ways it has been colorful, it has been remarkably monotonous in one respect. The central problem has remained unsolved throughout, and the arguments that created the major difficulty at the beginning are the very same that today still preclude any settlement of the question.

Towards the close of the 5th century B.C. the process of knowing had been conceptually framed in a relatively stable general scenario. By and large, the thinkers who concerned themselves with the cognizing activity tacitly accepted a scenario in which the knower and the things of which, or about which, he or she comes to know are, from the outset, separate and independent entities.

The problem arises from the "iconic" conception of knowledge, a conception that requires a *match* or *correspondence* between the cognitive structures and what these structures are supposed to *represent*. Truth, in that conception, inevitably becomes the perfect match, the flawless representation. The moment we accept that scenario, we begin to feel the need to assess just how well our cognitive structures match what they are intended to represent. But that "reality" lies forever on the other side of our experiential interface. To make any such assessment of truth we should have to be able, as Hilary Putnam recently put it, to adopt a "God's eye view" (Putnam,1982). Because we are not, and logically cannot be, in a position to have such a view of the "real" world and its presumed representation, there is no way out of the dilemma. What we need is a different scenario, a different conception of what it is "to know," a conception in which the goodness of knowledge is not predicated on likeness or representation.

The first explicit proposal of a different approach originated in those quarters that were most concerned with faith and its preservation. When, for the first time, the revolutionary notion that the Earth might not be the center of the universe seriously threatened the picture of the world that the church held to be unquestionable and sacred, it was the defenders of the faith who proposed an alternative scenario for the pursuit of scientific knowledge. In his preface to Copernicus' treatise *De revolutionibus,* Osiander (1627) suggested: "There is no need for these hypotheses to be true, or even to be at all like the truth; rather, one thing is sufficient for them—that they yield calculations which agree with the observations."[1]

This introduces the notion of a second kind of knowledge, apart from faith and dogma, a knowledge that *fits* observations. It is knowledge that human reason derives from experience. It does not represent a picture of the "real" world but provides structure and organization to experience. As such it has an all-important function: It enables us to solve experiential problems.

In Descartes' time, this instrumentalist theory of knowledge was formulated and developed by Mersenne and Gassendi.[2]

From an explorer who is condemned to seek "structural properties" of an inaccessible reality, the experiencing organism now turns into a builder of cognitive structures intended to solve such problems as the organism perceives or conceives. Fifty years ago, Piaget characterized this scenario as neatly as one could wish: "Intelligence organizes the world by organizing itself." (Piaget, 1937). What determines the value of the conceptual structures is their experiential adequacy, their goodness of *fit* with experience, their *viability* as means for the solving of problems, among which is, of course, the never-ending problem of consistent organization that we call *understanding.*

The world we live in, from the vantage point of this new perspective, is always and necessarily the world as we conceptualize it. "Facts," as Vico saw long ago, are *made* by us and our way of experiencing, rather than *given* by an independently existing objective world. But that does not mean that we can make

---

[1]Translation from Popper, K. R. (1968). *Conjectures and refutations* (p. 98). New York: Harper Torchbooks.

[2]An excellent exposition can be found in Popkin, R. H. (1979). *The history of scepticism from Erasmus to Spinoza.* Berkeley, CA: University of California Press.

them as we like. They are viable facts as long as they do not clash with experience, as long as they remain tenable in the sense that they continue to do what we expect them to do.

This view of knowledge, clearly, has serious consequences for our conceptualization of teaching and learning. Above all, it will shift the emphasis from the student's "correct" replication of what the teacher does, to the student's successful organization of his or her *own* experience. But before I expand on that I want to examine the widespread notion that knowledge is a commodity that can be communicated.

COMMUNICATION AND THE SUBJECTIVITY OF MEANING

The way we usually think of "meaning" is conditioned by centuries of written language. We are inclined to think of the meaning of words in a text rather than of the meaning a speaker intends when he or she is uttering linguistic sounds. Written language and printed texts have a physical persistence. They lie on our desks or can be taken from shelves, they can be handled and read. When we *understand* what we read, we gain the impression that we have "grasped" the meaning of the printed words, and we come to believe that this meaning was *in* the words and that we extracted it like kernels out of their shells. We may even say that a particular meaning is the "content" of a word or of a text. This notion of words as containers in which the writer or speaker "conveys" meaning to readers or listeners is extraordinarily strong and seems so natural that we are reluctant to question it. Yet, it is a misguided notion. To see this, we have to retrace our own steps and review how the meaning of words was acquired at the beginning of our linguistic career.

In order to attach any meaning to a word, a child must, first of all, learn to isolate that particular word as a recurrent sound pattern among the totality of available sensory signals. Next, she must isolate something else in her experiential field, something that recurs more or less regularly in conjunction with that sound pattern. Take an ordinary and relatively unproblematic word such as "apple." Let us assume that a child has come to recognize it as a recurrent item in her auditory experience. Let us further assume that the child already has a hunch that "apple" is the kind of sound pattern that *should* be associated with some other experiential item. Adults interested in the child's linguistic progress can, of course, help in that process of association by specific actions and reactions, and they will consider their "teaching" successful when the child has come to isolate in her experiential field something that enables her to respond in a way that they consider appropriate. When this has been achieved, when the appropriate association has been formed, there is yet another step the child must make before she can be said to have acquired the meaning of the word "apple."

The child must learn to re-present to herself the designated compound of experiences whenever the word is uttered, even when none of the elements of that compound are actually present in her experiential field. That is to say, the child must acquire the ability to imagine or visualize, for instance, what she has associated with the word "apple" whenever she hears the sound pattern of that word.[3]

This analysis, detailed though it may seem, is still nothing but a gross summary of certain indispensable steps in a long procedure of interactions. In the present context, however, it should suffice to justify the conclusion that the compound of experiential elements that constitutes the concept an individual has associated with a word cannot be anything but a compound of abstractions from that individual's own experience. For each one of us, then, the meaning of the word apple is an abstraction that he or she has made individually from whatever apple experiences he or she has had in the past. That is to say, it is subjective in origin and resides in a subject's head, not in the word that, because of an association, has the power to call up, in each of us, our own subjective representation.

If you grant this inherent subjectivity of concepts and, therefore, of meaning, you are immediately up against a serious problem. If the meanings of words are, indeed, our own subjective construction, how can we possibly communicate? How could anyone be confident that the representations called up in the mind of the listener are at all *like* the representations the speaker had in mind when he or she uttered the particular words? This question goes to the very heart of the problem of communication. Unfortunately the general conception of communication was derived from and shaped by the notion of words as containers of meaning. If that notion is inadequate, so must be the general conception of communication.

The trouble stems from the mistaken assumption that, in order to communicate, the representations associated with the words that are used must be the same for all communicators. For communication to be considered satisfactory and to lead to what we call "understanding," it is quite sufficient that the communicators' representations be compatible in the sense that they do not manifestly clash with the situational context or the speaker's expectations.

A simple example may help to make this clear. Let us assume that, for the first time, Jimmy hears the word "mermaid." He asks what it means and is told

---

[3]If this isolating of the named thing or "referent" is a demanding task with relatively simple perceptual compounds, such as apple, it is obviously much more difficult when the meaning of the word is a concept that requires further abstraction from sensory experience or from mental operations. But because we want to maintain that words such as "all" and "some," "mine" and "ours," "cause" and "effect," "space" and "time," and scores of others *have* meaning, we must assume that these meanings, though they cannot be directly perceived, are nevertheless somehow isolated and made retrievable by every learner of the language.

that a mermaid is a creature with a woman's head and torso and the tail of a fish. Jimmy need not have met such a creature in actual experience to imagine her. He can construct a representation out of familiar elements, provided he is somewhat familiar with and has established associations to "woman," "fish," and the other words used in the explanation. However, if Jimmy is not told that in mermaids the fish's tail replaces the woman's legs, he may construct a composite that is a fish-tailed biped and, therefore, rather unlike the intended creature of the seas. Jimmy might then read stories about mermaids and take part in conversations about them for quite some time *without* having to adjust his image. In fact, his deviant notion of a mermaid's physique could be corrected only if he got into a situation where the image of a creature with legs as well as a fish's tail comes into explicit conflict with a picture or with what speakers of the language say about mermaids; that is, Jimmy would modify the concept that is his subjective meaning of the word *only* if some context forced him to do so.

How, you may now ask, can a context *force* one to modify one's concepts? The question must be answered not only in a theory of communication but also in a theory of knowledge. The answer I am proposing is essentially the same in both.

The basic assumption is one that is familiar to you. Organisms live in a world of constraints. In order to survive, they must be "adapted" or, as I prefer to say, "viable." This means that they must be able to manage their living within the constraints of the world in which they live. This is commonplace in the context of biology and evolution. In my view, the principle applies also to cognition—with one important difference. On the biological level, we are concerned with species, i.e., with collections of organisms that, individually, cannot modify their biological makeup. But because they are not all the same, the species "adapts" simply because all those individuals that are *not* viable are eliminated and do not reproduce. On the cognitive level, we are concerned with individuals and specifically with their "knowledge," which, fortunately, is not immutable and only rarely fatal. The cognitive organism tries to make sense of experience in order better to avoid clashing with the world's constraints. It can actively modify ways and means to achieve greater viability.

"To make sense" is the same activity and involves the same presuppositions whether the stuff we want to make sense of is experience in general or the particular kind of experience we call communication. The procedure is the same but the motivation, the reason why we want to make sense, may be different.

Let me begin with ordinary experience. No matter how one characterizes cognizing organisms, one of their salient features is that they are capable of learning. Basically, to have "learned" means to have drawn conclusions from experience and to act accordingly. To act accordingly, of course, implies that there are certain experiences that one would like to repeat rather than others that one would like to avoid. The expectation that any such control over experience can be gained must be founded on the assumptions that (1) some regularities can be detected in the experiential sequence and (2) future experience will at least to some extent conform to these regularities. These assumptions, as David Hume showed, are prerequisites for the inductive process and the knowledge that results from it.

In order to find regularities, we must segment our experience into separate pieces so that, after certain operations of recall and comparison, we may say of some of them that they recur. The segmenting and recalling, the assessing of similarities, and the decisions as to what is to be considered different are all *our* doing. Yet, whenever some particular result of these activities turns out to be useful (in generating desirable or avoiding undesirable experiences), we quickly forget that we could have segmented, considered, and assessed otherwise. When a scheme has worked several times, we come to believe, as Piaget has remarked, that it could not be otherwise and that we have discovered something about the real world. Actually we have merely found *one* viable way of organizing our experience. "To make sense" of a given collection of experiences, then, means to have organized them in a way that permits us to make more or less reliable predictions. In fact, it is almost universally the case that we interpret experience either in view of expectations or with a view to making predictions about experiences that are to come.

In contrast, "to make sense" of a piece of language does not usually involve the prediction of future nonlinguistic experience. However, it does involve the forming of expectations concerning the remainder of the piece that we have not yet heard or read. These expectations concern words and concepts, not actions or other experiential events. The piece of language may, of course, be intended to express a prediction, e.g., "tomorrow it will rain," but the way in which that prediction is derived from the piece of language differs from the way in which it might be derived from, say, the observation of particular clouds in the sky. The difference comes out clearly when it is pointed out that, in order to make sense of the utterance "tomorrow it will rain" it is quite irrelevant whether or not there is any belief in the likelihood of rain. To "understand" the utterance, it is sufficient that we come up with a conceptual structure that, given our past experience with words and the way they are combined, *fits* the piece of language in hand. The fact that, when tomorrow comes, it doesn't rain, in no way invalidates the interpretation of the utterance. On the other hand, if the prediction made from an observation of the sky is not confirmed by actual rain, we have to conclude that there was something wrong with our interpretation of the clouds.

In spite of this difference between the interpretation of experience and the interpretation of language, the two have one important feature in common. Both rely on the use of conceptual material that the interpreter must already have. "Making sense," in both cases, means finding a way of fitting available conceptual elements into a pattern that is circumscribed by specific constraints. In the one case, the constraints are inherent in the way in which we have come to segment and organize experience; in the other, the constraints are inherent in the way in which we have learned to use language. In neither case is it a matter of matching an original. If our interpretation of experience allows us to achieve our purpose, we are quite satisfied that we "know"; and if our interpretation of a communication is not countermanded by anything the communicator says or does, we are quite satisfied that we have "understood."

The process of understanding in the context of communication is analogous to the process of coming to know in the context of experience. In both cases, it is a matter of building up, out of available elements, conceptual structures that fit into such space as is left unencumbered by constraints. Though this is, of course, a spatial metaphor, it illuminates the essential character of the notion of viability and it brings out another aspect that differentiates that notion from the traditional one of "truth": Having constructed a viable path of action, a viable solution to an experiential problem, or a viable interpretation of a piece of language, there is never any reason to believe that this construction is the only one possible.

THE CONSTRUCTION OF VIABLE KNOWLEDGE

When I began the section on communication by talking about the concept of meaning, it must have become apparent that I am not a behaviorist. For about half a century behaviorists have worked hard to do away with "mentalistic" notions such as *meaning, representation, and thought.* It is up to future historians to assess just how much damage this mindless fashion has wrought. Where education is concerned, the damage was formidable. Because behaviorism is by no means extinct, damage continues to be done, and it is done in many ways. One common root, however, is the presumption that all that matters—perhaps even all there is—are observable stimuli and observable responses. This presumption has been appallingly successful in wiping out the distinction between training and education.

As I hope to have shown in the preceding section, a child must learn more than just to respond "apple" to instantiations of actual apple experiences. If that were all she could do, her linguistic proficiency would remain equivalent to that of a well-trained parrot. For the bird and its trainer to have come so far is a remarkable achievement. For a human child it is a starting point in the development of self-regulation, awareness, and rational control.

As mathematics educators, you know this better than most. To give correct answers to questions within the range of the multiplication table is no doubt a useful accomplishment, but it is, in itself, no demonstration of *mathematical* knowledge. Mathematical knowledge cannot be reduced to a stock of retrievable "facts" but concerns the ability to compute new results. To use Piaget's terms, it is *operative* rather than *figurative*. It is the product of reflection—and whereas reflection as such is not observable, its products *may* be inferred from observable responses.

I am using "reflection" in the sense in which it was originally introduced by Locke, i.e., for the ability of the mind to observe its own operations. Operative knowledge, therefore, is not associative retrieval of a particular answer but rather knowledge of what to do in order to produce an answer. Operative knowledge is constructive and, consequently, is best demonstrated in situations where something new is generated, something that was not already available to the operator. The novelty that matters is, of course, novelty from the subject's point of view. An observer, experimenter, or teacher can infer this subjective novelty, not from the correctness of a response but from the struggle that led to it. It is not the particular response that matters but the way in which it was arrived at.

In the preceding pages, I have several times used the term *interpretation*. I have done it deliberately, because it focuses attention on an activity that requires awareness and deliberate choice. Although all the material that is used in the process of interpreting may have been shaped and prepared by prior interaction with experiential things and with people, and although the validation of any particular interpretation does, as we have seen, require further interaction, the process of interpreting itself requires reflection. If an organism does no more than act and react, it would be misusing the word to say that the organism is interpreting. Interpretation implies awareness of more than one possibility, deliberation, and rationally controlled choice.

A student's ability to carry out certain activities is never more than part of what we call "competence." The other part is the ability to monitor the activities. To do the right thing is not enough; to be competent, one must also know what one is doing and why it is right. That is perhaps the most stringent reason why longitudinal observation and Piaget's clinical method are indispensable if we want to find out anything about the reflective thought of children, about their operative knowledge, and about how to teach them to make progress towards competence.

At the beginning of this chapter, I mentioned that a useful analogy might be found in the teaching of athletic skills. What I was alluding to are the recently developed methods that make it possible for athletes to *see* what they are doing. Some of these methods involve stachistoscopy and are very sophisticated, others are as simple as the slow-motion replay of movies and videotapes. Their purpose is to give performers of intricate actions an opportunity to observe themselves act. This visual feedback is a far more powerful didactic tool than instructions that refer to details of the action that, normally, are dimly or not at all perceived by the actor himself.

The proficiency of good athletes springs to a large extent from the fact that they have, as it were, automated much of their action. As long as their way of acting is actually the most effective for the purpose, this automation is an advantage because it frees the conscious mind to focus on higher levels of control. When, however, something must be changed in the routine, this would be difficult, if not impossible, to achieve *without* awareness of the individual steps. Hence, the efficacy of visual feedback.

Although the speed of execution that comes with automation may be a characteristic of the expert calculator, the primary goal of mathematics instruction has to be the student's conscious understanding of what he or she is doing and why it is being done. This understanding is not unlike the self-awareness the athlete must acquire in order consciously to make an improvement in his physical routine. Unfortunately we have no stachistoscope or camera that could capture the dynamics, the detailed progression of steps, of the mental operations that lead to the solution of a numerical problem. Yet, what the mathematics teacher is striving to instill into the student is ultimately the awareness of a dynamic program and its execution—and that awareness is in principle similar to what the athlete is able to glean from a slow-motion re-presentation of his or her own performance. In the absence of any such technology to create self-reflection, the teacher must find other means to foster operative awareness. At the present state of the art, the method of the "teaching experiment" developed by Steffe seems to be the most hopeful step in that direction. (Steffe, 1977).

The term *teaching experiment* could easily be misunderstood. It is not intended to indicate an experiment in teaching an accepted way of operating, as for instance, the adult's way of adding and subtracting. Instead, it is primarily an exploratory tool, derived from Piaget's clinical interview and aimed at discovering what goes on in the student's head. To this it adds experimentation with ways and means of modifying the student's operating. The ways and means of bringing about such change are, in a sense, the opposite of what has become known as behavior modification.

A large part of educational research has been employing a procedure that consists of setting tasks, recording solutions, and analyzing these solutions as though they resulted from the child's fumbling efforts to carry out operations that constitute an adult's competence. The teaching experiment, instead, starts from the premise that the child cannot conceive of the task, the way to solve it, and the solution in terms other than those that are available at the particular point in the child's conceptual development. The child, to put it another way, must interpret the task and try to construct a solution by using material she already has. That material cannot be anything but the conceptual building blocks and operations that the particular child has assembled in her own prior experience.

Children, we must never forget, are not repositories for adult "knowledge" but organisms that, like all of us, are constantly trying to make sense of, to understand their experience. As Confrey (1980) noted:

> Most traditional measurements of student learning in mathematics assume that individual differences in students' concepts either vary insubstantially or are unimportant in their influences on the mathematics studied. . . . In contrast, if one assumes that there are a variety of ways of understanding a concept mathematically, individual differences in mathematics become the diversity in students' understandings of concepts or of mathematics itself. The clinical interview pro-

vides a means for searching for and exploring these individual understandings. (Confrey, 1980).

It is not in the least facetious to say that the interviewer's goal is to gain understanding of the child's understandings. The difference between the child interpreting (and trying to solve) a task in the given context, and the interviewer interpreting the child's responses and behavior in the context of the task, is that the interviewer can test his interpretation by deliberately modifying certain elements in the child's experiential field. The interviewer can also ask questions and see whether or not the responses are compatible with his or her conjectures about the child's conception of what is going on. Whenever an incompatibility crops up, the interviewer's conjectures have to be changed and their replacements tested again, until at last they remain viable in whatever situations the interviewer can think of and create.

In short, the interviewer is constructing a *model* of the child's notions and operations. Inevitably, that model will be constructed, not out of the child's conceptual elements, but out of conceptual elements that are the interviewer's own. It is in this context that the epistemological principle of *fit*, rather than *match*, is of crucial importance. Just as cognitive organisms can never compare their conceptual organizations of experience with the structure of an independent objective reality, so the interviewer, experimenter, or teacher can never compare the model he or she has constructed of a child's conceptualizations with what actually goes on in the child's head. In the one case as in the other, the best that can be achieved is a model that remains viable within the range of available experience.

The teaching experiment, as I suggested before, is, however, something more than a clinical interview. Whereas the interview aims at establishing "where the child is," the experiment aims at ways and means of "getting the child on." Having generated a viable model of the child's present concepts and operations, the experimenter hypothesizes pathways to guide the child's conceptualizations towards adult competence. In order to formulate any such hypothetical path, let alone implement it, the experimenter/teacher must not only have a model of the student's present conceptual structures but also an analytical model of the adult conceptualizations towards which his guidance is to lead.

The structure of mathematical concepts is still largely obscure.[4] This may seem a strange complaint, given the amount of work that has been done in the last 100 years to clarify and articulate the foundations of mathematics. As a result of that work there is no shortage of definitions, but these definitions, for the most part, are formal rather than conceptual; that is, they merely substitute other signs

---

[4]For recent conceptual analyses see Steffe, L. P., von Glasersfeld, E., Richards, J., & Cobb, P. (1983). *Children's counting types: Philosophy, theory, and application*. New York: Praeger.

or symbols for the definiendum. Rarely, if ever, is there a hint, let alone an indication, of what one must *do* in order to build up the conceptual structures that are to be associated with the symbols. Yet, that is of course what a child has to find out if he or she is to acquire a new concept.

Let me give you an example. A current definition of number, in the sense of "positive integer," says that it is "a symbol associated with a set and with all other sets which can be put into one-to-one correspondence with this set." (James & James, 1968). The mention of "put" makes it sound like an instruction to act, a directive for construction, which is what it ought to be. But, in order to begin that construction, the student would have to have a clear understanding of "set" and, more important still, of "one." Such understanding can be achieved only by reflecting on the operations of one's own mind and the realization that with these operations one can create units and sets anywhere and at any time, irrespective of any sensory signals. That means that, rather than speak of "sets" and "mathematical objects" as though they had an independent existence in some "objective" reality, teachers would have to foster the student's reflective awareness of his or her mental operations, because it is only from these that the required concepts can be abstracted.

The teaching experiment, at any rate, presupposes an explicit model of adult functioning. The experimental part of the method then consists in a form of "indirect guidance" aimed at modifying the child's present concepts and operations (which the experimenter "knows" in terms of the model constructed on the basis of observing the particular child) towards the adult concepts and operations (which the experimenter "knows" in terms of the model constructed on the basis of analyzing the adult procedures). Because the child necessarily interprets verbal instructions in terms of her own experience, the "guidance" must take the form either of questions or of changes in the experiential field that leads the child into situations where her present way of operating runs into obstacles and contradiction. Analogous to the adult who organizes general experience, the child is unlikely to modify a conceptual structure unless there is an experience of failure or, at least, surprise at something not working out in the expected fashion. Such failure or surprise, however, can be experienced only if there was an expectation—and that brings me to the last point I want to make.

If I have had any success at all in presenting the constructivist epistemology as a possible basis for education and educational research, this last point is easy to make and its importance should become obvious.

The more abstract the concepts and operations that are to be constituted, the more reflective activity will be needed. Reflection, however, does not happen without effort. The concepts and operations involved in mathematics are not merely abstractions, but most of them are the product of several levels of abstraction. Hence, it is not just one act of reflection that is needed, but a succession of reflective efforts—and any succession of efforts requires solid motivation.

The need for motivation is certainly no news to anyone who has been teaching. How to foster motivation has been discussed for a long time. But here again, I believe, the effect of behaviorism has been profoundly detrimental. The basic dogma of behaviorism merely says that behavior is determined by the consequences it has produced in the past (which is just another way of saying that organisms operate inductively). There is every reason to agree with that. The trouble arises from the usual interpretation of ''reinforcement,'' i.e., of the consequence that is rewarding and thus strengthens specific behaviors and increases the probability of their recurrence.

There is the widespread misconception that reinforcement is the effect of certain well-known commodities such as cookies, money, and social approval. It is a misconception, not because organisms will not work quite hard to obtain these commodities, but because it obscures the one thing that is often by far the most reinforcing for a cognitive organism: *to achieve a satisfactory organization,* a viable way of dealing with some sector of experience. This fact adds a different dimension to the conception of reinforcement because whatever constitutes the rewarding consequence in these cases is generated wholly *within* the organism's own system.

Self-generated reinforcement has an enormous potential in cognitive, reflective organisms. (All of us, I am sure, have spent precious time and sweat on puzzles whose solution brought neither cookies, nor money, and negligible social approval.) But this potential has to be developed and realized.

When children begin to play with wooden blocks, they sooner or later place one upon another. Whatever satisfaction they derive from the resulting structure, it provides sufficient incentive for them to repeat the act and to improve on it. They may, for instance, implicitly or explicitly set themselves the goal of building a tower that comprises *all* the blocks. If they succeed, they are manifestly satisfied, irrespective of tangible rewards or an adult's comment, for they build towers also in the absence of observers. The reward springs from the achievement, from the successful deliberate imposition of an order that is inherent in their own way of organizing. To repeat the feat, the tower has to be knocked down. That, too, turns out to be a source of satisfaction because it once more provides evidence of the experiencer's power over the structure of experience.

To some, these observations may seem trivial. To me, they exemplify a basic feature of the model of the cognitive organism, a feature that must be taken into account if we want to educate.

From the constructivist point of view, it makes no sense to assume that any powerful cognitive satisfaction springs from simply being told that one has done something right, as long as ''rightness'' is assessed by someone else. To become a source of real satisfaction, ''rightness'' must be seen as the fit with an order one has established *oneself.* Teachers as well as mathematicians tend to assume that there exists in every particular case an objective problem and an objectively "true" solution. Children and students of any age are therefore expected somehow to come to "see" the problem, its solution, and the *necessity* that links the two. But the necessity is conceptual and it can spring from nothing but the awareness of the structures and operations involved in the thinking subject's conceptualization of the problem and its solution. Logical or mathematical necessity does not reside in any independent world—to see it and gain satisfaction from it, one must reflect on one's own constructs and the way in which one has put them together.

FINAL REMARKS

Educators share the goal of generating knowledge in their students. However, from the epistemological perspective I have outlined, it appears that knowledge is not a transferable commodity and communication not a conveyance.

If, then, we come to see knowledge and competence as products of the individual's conceptual organization of the individual's experience, the teacher's role will no longer be to dispense "truth" but rather to help and guide the student in the conceptual organization of certain areas of experience. Two things are required for the teacher to do this: on the one hand, an adequate idea of where the student is and, on the other, an adequate idea of the destination. Neither is accessible to direct observation. What the student says and does can be interpreted in terms of a hypothetical model—and this is one area of educational research that every *good* teacher since Socrates has done intuitively. Today, we are a good deal closer to providing the teacher with a set of relatively reliable diagnostic tools.

As for the helping and guiding, good teachers have always found ways and means of doing it because, consciously or unconsciously they realized that, although one can point the way with words and symbols, it is the student who has to do the conceptualizing and the operating.

That leaves the destination, the way of operating that would be considered "right" from the expert's point of view. As I have mentioned earlier, a conceptual model of the formation of the structures and the operations that constitute mathematical competence is essential because it, alone, could indicate the direction in which the student is to be guided. The kind of analysis, however, that would yield a step by step path for the construction of mathematical concepts has barely been begun. It is in this area that, in my view, research could make advances that would immediately benefit educational practice. If the goal of the teacher's guidance is to generate understanding, rather than train specific performance, his task will clearly be greatly facilitated if that goal can be represented by an explicit model of the concepts and operations that we assume to be the operative source of mathematical competence. More important still, if students are to taste something of the mathematician's satisfaction in doing mathematics, they cannot be expected to find it in whatever rewards they might be given for their performance but only through becoming aware of the neatness of fit they have achieved in their own conceptual construction.

REFERENCES

Cobb, P. & Steffe, L. P. (1983). The constructivist researcher as teacher and model builder. *J.R.M.E.*, *14*, pp. 34–94.

Confrey, J. (1980). Clinical interviewing: Its potential to reveal insights in mathematical education. In R. Karplus (ed.) *Proceedings of the 4th International Conference for the Psychology of Mathematics Education*. Berkeley, CA.

James & James. (1968). *Mathematical dictionary*, (3rd edition). Princeton, NJ: Van Nostrand.

Piaget. J. (1937). *La construction du réel chez l'enfant*. Neuchâtel: Delachaux et Niestlé.

Popkin, R. H. (1979). *The History of Scepticism from Erasmus to Spinoza*. Berkeley, CA: University of California Press.

Popper, K. R. (1968). *Conjectures and refutations*. New York: Harper Torchbooks.

Putnam, H. (1982). *Reason, truth, and history*. Cambridge, MA: Harvard University Press.

Steffe, L. P. (1977). *Constructivist models for childrens' learning in arithmetic*. Paper presented at the Research Workshop on Learning Models, Durham, NH.

Steffe, L. P., von Glaserfeld, E., Richards, J., and Cobb, P. (1983). *Childrens' counting types: Philosophy, theory, and application*. New York: Praeger.

2 Representation Systems and Mathematics

James J. Kaput
Southeastern Massachusetts University

This chapter is one of a series of examinations dealing with the idea of representation in mathematics education. They are motivated by a tandem of practical concerns and theoretical needs. The practical concerns center on the well-known, deep, and continuing difficulties experienced by students in translating between different representations of mathematical ideas, and between common experience and mathematics. These concerns are magnified by the radical new range of representational opportunities offered by information technologies and the heightened requirement for systematic knowledge regarding the ways that such representations function and interact (Kaput, in press). The theoretical needs are a consequence of these practical matters coupled with the apparent lack of a comprehensive, systematic theoretical framework of symbol systems and representation systems capable of supporting the kinds of understandings necessary to solve the problems just alluded to. The theoretical perspective can be thought of as providing a linguistic/semiotic complement to the purely cognitive perspective on these problems.

Some mathematics education researchers have, in response to this need for understanding forms of representation in a particular area, developed local theories, e.g., to account for the differential ability of certain representations of rational numbers to carry certain aspects of the arithmetic of rational numbers better than other representations (Behr, Wachsmuth, Post, & Lesh, 1984; Chaffe-Stengel & Noddings, 1982; Lesh, Hamilton, Post, & Behr, 1983; Post, Wachsmuth, Lesh, & Behr, 1985). There have also been sporadic efforts to address the issues more broadly, the most successful example of which is probably the collection edited by Skemp (1982). However, a coherent and unifying theoretical context is lacking. This lack may be of critical importance to the efficacy of empirical work and its ultimate impact on the curriculum and instruction (Kilpatrick, 1981). Others have devoted considerable effort to the study of symbol systems outside of mathematics (Bruner, 1973; Cassirer, 1950; Gardner, 1978; Gardner, Howard, & Perkins, 1974; Goodman, 1968; Langer, 1942; Salomon, 1979; among others). Although some of these works are classics and monumental in scope, they have little to say directly about the issues that would inform the mathematics curriculum.

The overarching assumption of the line of inquiry advocated in this chapter and Chapter 14 is that more direct and systematic attention needs to be paid to the ways we use symbols and combinations of symbols in mathematical representation systems and the ways these systems relate to one another. An adequate theory ultimately would define general properties of symbol and representational systems themselves, as opposed to their potentially variable informational content. It would provide systematic criteria and vocabulary for discussing how the different representation systems vary in their ability to encode certain mathematical ideas and support the solving of certain classes of problems. These would be characterized in terms of psychologically identifiable structures and processes, such as the amount or kind of processing required to extract certain information from a particular encoding or to translate a given piece of information from one system to another. The theory eventually would provide a developmental trajectory for representational skills that would interface with descriptions of the other symbolic competences synchronically developing within the student, especially natural language competences (Gardner, 1978; 1983; Nesher & Schwartz, 1982). And it would necessarily connect with the existing corpus of knowledge in mathematics education.

To construct and support such a theory will not be an easy task and will surely take a number of years. We will outline a strategy for beginning the effort. But first we would like to be a bit more specific regarding the types of questions that this kind of theory might illuminate.

REPRESENTATION IS UNDERREPRESENTED

There is a common tendency to underestimate the role of representation systems in the standard curriculum. For example, we usually assume the mathematics curriculum in the first 8 years of school is about numbers, whereas the actual school work is mainly about a particular representation system for numbers—the base 10 placeholder system—and *its* properties, and the representational systems for rational numbers. The essence and power of numerical algorithms reside in the freedom to deal only with the *representations* of numbers without regard to the numbers they represent. Because of the construction of the representation system and the design of the algorithm, we can then be confident that the symbol produced by the symbol manipulation algorithm represents the actual answer.

Some perennial difficulties get seated in a different context when seen in these terms. For example, the failure of students to estimate or maintain an order of magnitude sense of a calculation (Davis & McKnight, 1980) can in fact be seen as a failure to cross between symbol and referent. There are also aspects of the skill automaticity or ''rote vs. meaningful learning'' question that can be stated in these terms (Wachsmuth, 1982, 1983) as well as aspects of the procedural vs conceptual knowledge issue.

The curriculum likewise ignores what may in some instances be an important distinction between those properties of numbers that are sensitive to the representation system versus those properties that are relatively independent of the representation system, e.g., primeness of numbers. We give virtually no explicit attention at any level in mathematics education to the relation between the transparency of certain mathematical properties or operations and the representation in which they are encoded. For example, the Cantor ''excluded middle'' set is very transparently constructed using the base three representation of real numbers, and some of its properties likewise are easily deduced from this representation, but not from others. Sensitivity to choice of representation is an old story in problem solving. (Recall the classic question regarding the monk who climbs to a mountaintop in one 12-hour span, prays 12 hours, and then spends another 12 hours descending on the same path he used for the ascent. Is there a place on the mountain where he is at exactly the same place descending at exactly the same time of day as when he was ascending 24 hours earlier?) Similarly, we omit from explicit instruction why some representations carry certain quantitative relationships more efficiently than others, as with rectangular versus polar coordinates. Likewise, we seldom deal with the critical issue of scaling when graphing, especially the relation between the choice of appropriate nonlinear scales and the simplicity of the graphical representation of specific quantitative relationships, e.g., logarithm scales and exponential growth/decay.

Naturally, there is a tight connection between the omissions of the curriculum and the omissions of the research community. However, as mentioned earlier the increasing availability of computer representations will magnify the importance of these omissions because instructors will have an enormous new variety of representations of mathematical ideas to work with, some of which may support one another, others of which may interfere with one another. The students of the near future likewise will be choosing how to represent given relationships. This skill in choosing or building representations, together with interpretive skills, will likely soon outstrip computational skills in importance by a wide margin. Both teachers and students will need more than uninformed intuition to guide them. They will need systematic, reliable, and psychologically valid knowledge about representations.

Another relevant problem area is translation between nonmathematical representations and mathematical ones. As noted in Kaput and Sims-Knight (1983) and elsewhere, student difficulty in such translation tasks is legend, yet very little focal attention is given to the teaching of how the various mathematical and nonmathematical representation systems work and how they relate to one another. On the other hand research indicates that this is a critical component in the ability to translate between systems (Sims-Knight & Kaput, 1983). A good theory of representation and symbol systems would be able to characterize in systematic ways the failure of systems to fit together smoothly. This is reflected in the "Students–Professors Problem" reversal error, where certain (multiplicative) quantitative relationships are expressed in mutually opposing ways in natural language versus algebra (Clement, 1982; Kaput & Sims-Knight, 1983). Similarly, certain representations are "cognitively orthogonal" when given pictorially versus graphically, as with the picture of a cyclist traveling over a hill versus the graph of her speed as a function of time (Sims-Knight & Kaput, 1983). We need more coherent and comprehensive ways of thinking about such phenomena.

STRATEGY FOR A THEORY

We now turn to the strategy for constructing the theoretical framework advocated earlier. The fundamental premise is that the root phenomena of mathematics learning and application are concerned with representation and symbolization because these are at the heart of the content of mathematics and are simultaneously at the heart of the cognitions associated with mathematical activity.

Our working premise is that the hardest part of the theory development will be establishing the psychological/linguistic reality of the constructs associated with external symbol creation and its bonding with internal cognitive structures and processes. The opaqueness of this bonding suggests that it be treated temporarily as a black box until enough conceptual and methodological tools are in hand to open it. Thus we begin informally, first with an examination of the content of mathematics itself from the representational perspective without regard to psychological, linguistic, or curricular constructs. We then take a first step in the direction of a psychological and linguistic theory by devising a quasiformal notion of representational system that makes sense in terms of the existing core mathematics curriculum. We show how basic mathematical ideas relate when viewed from this perspective and discuss the various characteristics of these systems, especially with regard to their symbol systems, which play a critical functional role. The third stage in the evolution of the theory is an explication of the quasiformal notion of representation system in terms of basic psychological phenomena connected with the representation of root mathematical concepts such as number and variable, and the role of the ambient representation systems of natural language and mental imagery. Subsequent steps examine how symbol systems build on one another and how intersubjectively shared external symbols interact actively with their internal referents.

DEFINING "REPRESENTATION"

There is real danger in stretching an undefined term such as *representation* across domains as different as formal mathematics, cognition, and epistemology. We deal with this danger by defining the term more precisely in several stages of increasing specificity and formality. We begin with a proto-definition that serves as a definitional framework into which more specific definitions can be fit later.

We follow Palmer (1977), and in fact many others, in assuming that any concept of representation must involve two related but functionally separate entities. We call one entity the representing world and the other the represented world. There should then be a correspondence between some aspects of the represented world and some aspects of the representing world. Hence any particular specification of a representation should describe the following five entities: (1) the represented world, (2) the representing world, (3) what aspects of the represented world are being represented, (4) what aspects of the representing world are doing the representing, and (5) the correspondence between the two worlds.

In many of the interesting cases one or both of the worlds may be hypothetical entities or even abstractions. There are four broad and interacting types of representation that can be accommodated under the preceding general framework: (1) cognitive and perceptual representation (Palmer's focus), (2) explanatory representation involving models, (3) representation within mathematics, and (4) external symbolic representation.

Our strategy for theory development can be viewed in part as aiming for integration of all four types of representation. In Kaput (1985) we introduced the definition of representation and dealt with (2) in some detail, focusing on computer models of mental representations and the question of explanatory adequacy. We now sketch briefly how the content of mathematics can be viewed as formal representation (3). The next step in our inquiry, a description of portions of the core school mathematics curriculum in terms of representation systems, appears in chapter 14 later in this volume. There the broad notion of representation system will be narrowed to the notion of *symbol system*.

REPRESENTATION IN MATHEMATICS

Mathematics proper may be regarded as the science of significant structure. Thus mathematics studies the representation of one structure by another, and much of the actual work of mathematics is to determine exactly what structure is preserved in that representation. This representation is formally independent of the form of the external symbols used because the structure itself is treated as an abstraction or idealization. Of course, in practice the material symbols do play a crucial role at every level of mathematical representation as suggested earlier.

We now outline several of the more common kinds of representation in mathematics.

Morphisms

The most frequently used construction in all mathematics involves structure-preserving mappings from one mathematical structure to another of the same general type. For example, a homomorphism $f: A \rightarrow B$ between groups is a representation in the sense defined earlier, where the mapping is the correspondence and, depending on the particular application at hand, either A or B can be the represented world. The aspects of each involved in the representation are the respective group structures as embodied in the operations and the rules they obey. These can be entirely abstract, or in other cases relatively concrete in the sense of being a group whose elements and operation are a familiar construct. For example, B may be a group of matrices with the operation of matrix addition. There are whole branches of mathematics based on the development and application of such representations, e.g., the representation theory of finite groups. An elementary example of such is given by Cayley's Theorem, which states that any group is isomorphic to a group of permutations on the set of its elements. This example illustrates a stronger form of representation wherein the correspondence is an embedding.

Generic Algebraic Constructions

Most of the standard constructions of abstract algebra can be regarded as providing a simplified representation of a given abstract object, as with the quotient of a group by a (normal) subgroup, which provides a simplified version of the original group with all the complexity of the subgroup "factored out." A similar remark applies to the role of Fundamental Homomorphism Theorems for algebraic objects of most any sort.

Canonical Building-Block Constructions (External)

Frequently an object A is represented by B via an isomorphism $f: A \rightarrow B$ where B is constructed from simpler structures of the given type. For example The Fundamental Theorem of Finite Abelian Groups asserts an isomorphism between any finite commutative group and a direct product of cyclic groups of integers.

Canonical Building-Block Constructions (Internal)

Some representations are defined in a more "internal" fashion, as in the Fundamental Theorem of Arithmetic (any integer has a factorization into a product of primes in an essentially unique way). Here the representation is not defined by a structure-preserving mapping but by an equation. As with earlier examples, the Fundamental Theorem of Algebra (every polynomial has a factorization into linear factors in an essentially unique way), and a whole host of similar types of constructions, a given object is represented by simpler, more atomistic objects.

Approximation

A class of representations occurring all across mathematics involves approximation of a given object by another simpler or more primitive object, as with the Taylor Polynomial, or series approximations of suitably differentiable functions by polynomial functions or series. The correspondence is the approximation itself. Much of classical analysis revolves around such representations.

Feature/Property Isolation

Another class of representations disambiguates a particular aspect of a mathematical object from the other features of the object by creating a new object of a different kind that represents that aspect cleanly. For example, the Fundamental Group of a topological space describes the ''holes'' (simply-connectedness) of the space in a precise way. Other examples of such (functorial) representation assignments abound.

Logic Models

A slightly different kind of representation is used in classic logic, where, given an axiom system with a collection of undefined primitives, one makes a formal interpretation of the primitives in some other, presumably more familiar system S that serves as a representation (''model'') of the axiom system. In this way axioms are tested for consistency, categoricity, dependence, etc.

CONCLUSION

The frequent use of the word ''fundamental'' in the previous discussion is not an accident. Most of the results that mathematicians regard as truly fundamental are easily classifiable as representational. Also, much actual research in mathematics is an attempt to extend such representations to new mathematical domains involving more esoteric objects. The representational goals remain the same.

It should be apparent that *the idea of representation is continuous with mathematics itself.*

Having established the representational character of mathematics at an abstract level, our next step is to study the varieties of materially realizable forms, and the psychological aspects of those forms, by which the representations are achieved.

REFERENCES

Behr, M., Wachsmuth, I., Post, T., & Lesh, R. (1984). Order and equivalence of rational numbers: A clinical teaching experiment. *Journal for Research in Mathematics Education, 15,* 323–341.

Bruner, J. (1973). In J. Anglin (ed.), *Beyond the information given.* New York: Norton.

Cassirer, E. (1950). *The problem of knowledge.* New Haven, CT: Yale University Press.

Chaffe-Stengel, P., & Noddings, N. (1982). Facilitating symbolic understanding of fractions. *For the Learning of Mathematics, 3* (2).

Clement, J. (1982). Algebra word problem solutions: Thought processes underlying a common misconception. *Journal for Research in Mathematics Education, 13,* 16–30.

Davis, R. B., & McKnight, C. (1980). The influence of semantic content on algorithmic behavior. *Journal of Mathematical Behavior, 3* (1).

Gardner, H. (1978). *Developmental psychology after Piaget: An approach in terms of symbol systems.* Unpublished manuscript, Harvard Project Zero, Harvard University, Cambridge, MA.

Gardner, H. (1983). Frames of mind: The theory of multiple intelligences. New York: Basic Books.

Gardner, H., Howard, V., & Perkins, D. (1974). Symbol systems: A philosophical, psychological, and educational investigation. In D. Olson (Ed.), *Media and symbols: The forms of expression, communication, and education.* Chicago: University of Chicago Press.

Goodman, N. (1968). *Languages of art.* Indianapolis: Bobbs–Merrill.

Kaput, J. (1985). Representation and problem solving: Some methodological issues. In E. Silver (Ed.), *Teaching and learning mathematical problem solving: Multiple research perspectives.* Hillsdale, NJ: Lawrence Erlbaum Associates.

Kaput, J. (in press). Information technology and mathematics: Opening new representational windows. *Journal of Mathematical Behavior.*

Kaput, J., & Sims-Knight, J. (1984). Errors in translations to algebraic equations: Roots and implications. In M. Behr & G. Bright (Eds.), Special issue on mathematics learning problems of the postsecondary student. *Focus on learning problems in mathematics.*

Kilpatrick, J. (1981). The reasonable ineffectiveness of research in mathematics education. *For the Learning of Mathematics, 2* (2).

Langer, S. (1942). *Philosophy in a new key.* Cambridge, MA: Harvard University Press.

Lesh, R., Hamilton, E., Post, T., & Behr, M. (1983). *Rational number relations and proportions.* (unpublished manuscript).

Nesher, P., & Schwartz, J. (1982, March). *Early quantification.* Paper given at MIT Division for Research and Science Education, Cambridge, MA.

Palmer, S. E. (1977). Fundamental aspects of cognitive representation. In E. Rosch & B. B. Lloyd (Eds.), *Cognition and categorization.* Hillsdale, NJ: Lawrence Erlbaum Associates.

Post, T., Wachsmuth, I., Lesh, R., & Behr, M. (1985). Order and equivalence of rational numbers: A cognitive analysis. *Journal for Research in Mathematics Education, 16,* 18–36.

Salomon, G. (1979). *Interaction of media, cognition, and learning.* San Francisco: Jossey–Bass.

Sims-Knight, J., & Kaput, J. (1983). Misconceptions of algebraic symbols: Representations and component processes. In J. Novak (Ed.), *Proceedings of the International Seminar on Misconceptions in Mathematics and Science* (pp. 477–488). Columbus, OH: ERIC Clearinghouse for Science, Mathematics, and Environmental Education.

Skemp, R. (1982). Understanding the symbolism of mathematics. Special issue of *Visible language, 16* (3).

Wachsmuth, I. (1982). Letter to the editor. *Journal of Mathematical Behavior, 3* (2).

Wachsmuth, I. (1983). Skill automaticity in mathematics instruction: A response to Gagné. *Journal for research in mathematics education, 14,* 204–209.

3 Translation Processes in Mathematics Education

Claude Janvier
Université du Québec à Montréal, Canada

In addition to the usual (but too often overlooked) geometrical figures that in many respects are abstract, "the repertoire" of modes of representation is extensive: Venn diagrams, tree diagrams, Cayley tables, arrow graphs, flow diagrams, cartesian graphs . . . as well as letters, numerals in various positions, brackets, . . . (Bell, A. W., 1976). Even from a pedagogical point of view, most textbooks today make use of a wide (if not wild) variety of diagrams and pictures meant to promote understanding. Everyone certainly agrees that the use of symbolism in mathematics thinking is fundamental.

However, particular uses of symbolism appear generally to be overlooked, namely, the translation processes. By a translation process, we mean the psychological processes involved in going from one mode of representation to another, for example, from an equation to a graph.

In recent years, Lesh (1979) points their importance in real problem solving, and Burton (1979) in an inventory of problem-solving skills mentions the notion of translation in her list. Burkhardt[1] (1977) stresses their crucial role in mathematical modeling. M. Bell (1979) identifies this fundamental component of mathematical competence. Bessot and Richard (1979) conducted a research on tree diagrams and graphs. It is not easy to repertoriate the literature dealing with translation processes because quite often the idea is tackled indirectly. For instance, researchers concerned with the notions of models, intuitive models (for example, see Fischbein, 1977), implicitly touch upon this question.

We personally got interested in this topic in a recent research project dealing with the "interpretation of cartesian graphs representing situations" (Janvier,

[1]To whom we extend our thanks for the idea.

1978). We have selected seven points on which we briefly report our findings in this chapter. During our talk, we illustrate most points with examples and we invite the participants to share their experience.

The Question of Giving Names

If we limit the modes of representation of variables to four, namely, verbal description, table, graph, and formula (equation), we can represent translation processes in the context of handling variables by a 4×4 table.

TRANSLATION PROCESSES

From \ To	Situations, Verbal Description	Tables	Graphs	Formulae
Situations, Verbal Description		Measuring	Sketching	Modelling
Tables	Reading		Plotting	Fitting
Graphs	Interpretation	Reading off		Curve fitting
Formulae	Parameter Recognition	Computing	Sketching	

The reader most probably feels a little uneasy about how a few cells were filled in, or about the names given to a few processes. He is kindly invited to fill those cells from scratch and discover the source of his hesitation. In fact, the quasi-impossibility of defining (at least uniquely) a few processes arises from the ill-defined context in which a particular translation is achieved. Despite this limitation, this array is very useful in that it gives a total picture of the question and helps us to single out and illustrate a variety of its aspects.

The reader has noted that we did not fill the diagonal. In fact, such processes exist that we could call transposition.

Direct or Indirect?

Let us simply look at a subtable for the time being (setting aside the verbal description, sometimes of situation).

TRANSLATION PROCESSES

To / From	Situations, Verbal Description	Tables	Graphs	Formulae
Situations, Verbal Description		Measuring	Sketching	Modelling
Tables	Reading		Plotting	Fitting
Graphs	Interpretation	Reading off		Curve fitting
Formulae	Parameter Recognition	Computing	Sketching	

As you notice, a few arrows are now added to account for alternative ways to achieve translations. More precisely, these arrows indicate that the translation "table → formula" is often carried out as "table → graph → formula" and "formula → graph" as "formula → table → graph." Actually indirect processes are substantially different than direct ones. A study of several mathematics programs has showed that they exclusively develop the indirect version of many processes of the tables. Could we do otherwise? Would it be worthwhile?

The Source and Target Paradigm

A translation involves two modes of representation. For the modes equation and graph, for instance, we have the translations: "graph → equation" and "equation → graph." To achieve directly (and correctly) a given translation, one has to transform the source "target-wise" or, in other words, to look at it from a "target point of view" and derive the results. For instance a graph has to be examined equation-wise (with respect to its equational features). This simple comment suggests teaching strategies as well as points to the importance of the inverse process (which we prefer to call complementary), which is symmetrically located with respect to the diagonal. In fact, our research suggested that processes were best developed in (symmetric) pairs, in our case, "graph → verbal description" (interpretation) and "verbal description → graph" (sketching). This principle is already applied by those teaching methods of foreign language that stress "listening" and "talking" in language laboratory.

The Role of Language

We have just stated that the source has to be examined "target-wise" but our research results equally show that in the process language or words play a central role. Actually, we noted that verbal tags were given to the relevant elements and the execution of the process was carried out through an efficient handling of those verbal tags. In a way, source and target were verbally simplified. Even in a "graphical context," the speech appeared to be essential.

Application to the Teaching of Music

Clearly, this type of analysis can be applied to other areas of mathematics (proportion, arithmetic operations) and other domains of education. We take very easy examples of a process involved in music, namely some related to a single note. We distinguish four modes of representation of a single note: the sound (vibration in the air), the act of playing it (position of finger(s), body), the name of the note (ex.: A or La), and its written symbol on the staff. We can consider then all the questions posed so far on the following table.

TRANSLATION PROCESSES.

From \ To	Vibrating note	Played note (action)	Name of note	Symbol on staff
Vibrating note		Play by ear	Tell name of a note	
Played note (action)				
Name of note	Sing			
Symbol on staff	Sing		Read music	

a) You see that we have not given names to all cells!

b) Play by ear: "vibrating note (heard) → played note (action)" is a direct process. For many, the process is often impossible or sometimes achieved indirectly: "vibrating note → name of a note → played note."

c) Traditional teaching method stresses the process "name of note → played note'' whereas the Suzuki method is based on the process ''vibrating (heard) note → played note (action).''

d) Look at how inverse processes are interrelated!

More Complex Thinking Processes in Mathematics Education

This section states opinions rather than research observations. We believe that the study of generalization, abstraction, proof, symbolization . . . could be made more meaningful if they were considered as made out of basic translation processes.

The study of intuition should equally gain from using the ''translation process'' framework. We believe with Fischbein that intuitions are based on schemata, verbal descriptions . . . which are implicitly used in action. We think that the development of intuition is often related to the development of processes that are subsumed and later called into play.

Translation Processes and Curriculum Designs

Using the idea of translation processes involving variables, we carried out in our doctoral dissertation (Janvier, 1978) a comparison between two mathematics programs that proved most interesting for the improvements it suggested. Translation processes for other topics could be used fruitfully.

CODA

Lately, we have been studying the question from different points of view. To answer the question, ''Is there anything behind all those modes of representation,'' we are inquiring about answers semiology (science of meaning) could give. Also, because each mode has its intrinsic level of complexity, we wonder how the theory of information and the idea of entropy could help out.

REFERENCES

Bell, A. W. (1976). *The learning of general mathematical strategies.* Doctoral dissertation. Shell Centre for Mathematics Education, University of Nottingham.

Bell, M. (1979). *What most people need from mathematics: Teaching mathematics so as to be useful.* Unpublished paper.

Bessot, A., & Richard, F. (1979). *Commande des variables dans une situation didactique pour provoquer l'élargissement de procédures en vue d'étudier le rôle du schéma.* Thèse collective de 3e cycle, Université de Bordeaux I.

Burkhardt, H. (1977). *Seven sevens are fifty? Mathematics for the real world.* Inaugural Lecture. Shell Centre for Mathematics & Education Publication, University of Nottingham.

Burton, L. (1979). The classification of problem-solving skills and procedures—presentation of an inventory. *Proceedings of the 3rd International Conference of IGPME,* Warwick.

Fischbein, E. (1977). Image and concept in learning mathematics. *Educational Studies in Mathematics, 8,* 153–165.

Janvier, C. (1978). *The interpretation of complex cartesian graph representing situations—studies and teaching experiments.* Doctoral dissertation. University of Nottingham.

Lesh, R. (1979). Some trends in research and the acquisition and the use of space and geometry concepts. *Critical Reviews in Mathematics Education,* Materialen and Studien, Band 9. Institut für Didactik der Mathematik der Universität Bielefeld.

4 Representations and Translations among Representations in Mathematics Learning and Problem Solving

Richard Lesh
WICAT & Northwestern University

Tom Post
University of Minnesota

Merlyn Behr
Northern Illinois University

This chapter briefly describes several roles that representations, and translations among representations, play in mathematical learning and problem solving. The term *representations* here is interpreted in a naive and restricted sense as external (and therefore observable) embodiments of students' internal conceptualizations—although this external/internal dichotomy is artificial.

Comments in this chapter are based on three recent or current National Science Foundation funded projects on Applied Mathematical Problem Solving (AMPS), Proportional Reasoning (PR), and Rational Number (RN) concept formation. Past PN, PR, and AMPS publications (e.g., Behr, Lesh, Post, & Silver, 1983; Lesh, 1981; Lesh, Landau, & Hamilton, 1983) have identified five distinct types of representation systems that occur in mathematics learning and problem solving (see Fig. 4.1); they are: (1) experience-based "scripts"—in which knowledge is organized around "real world" events that serve as general contexts for interpreting and solving other kinds of problem situations; (2) manipulatable models—like Cuisenaire rods, arithmetic blocks, fraction bars, number lines, etc., in which the "elements" in the system have little meaning per se, but the "built in" relationships and operations fit many everyday situations; (3) pictures or diagrams—static figural models that, like manipulatable models, can be internalized as "images"; (4) spoken languages—including specialized sublanguages related to domains like logic, etc.; (5) written sym-

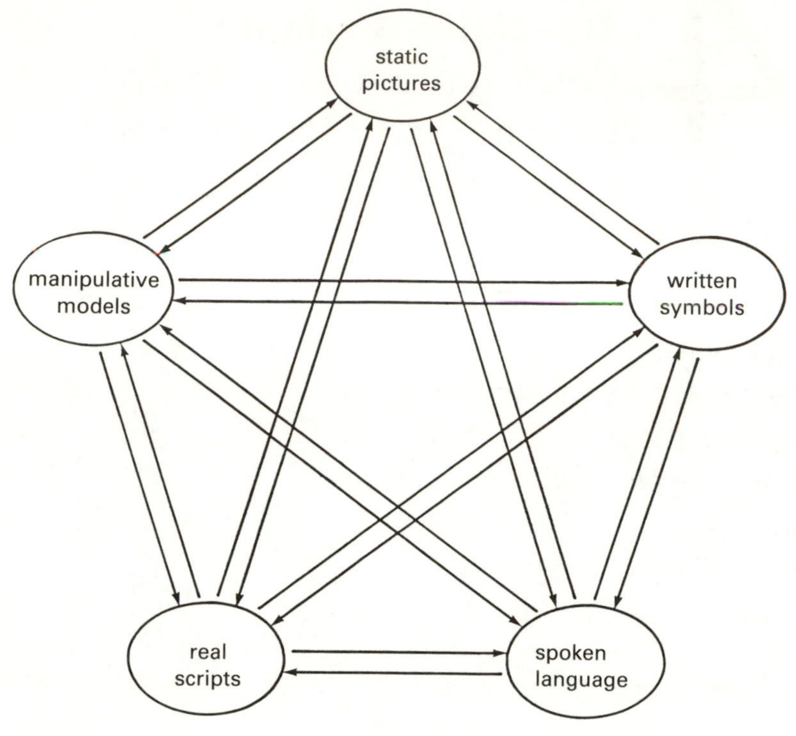

FIG. 4.1.

bols—which, like spoken languages, can involve specialized sentences and phrases ($X + 3 = 7$, $A' \cup B' = (A \cap B)'$) as well as normal English sentences and phrases.

This chapter emphasizes that, not only are these distinct types of representation systems important in their own rights, but *translations* among them, and *transformations* within them, also are important.

Item 31 (Fig. 4.2), taken from a written test on "rational number relations and proportions" from our RN/PR projects, illustrates a "written symbol to picture" translation. The aim is to require students to answer the item correctly by establishing a relationship (or mapping) from one representational system to another, preserving structural characteristics and meaning in much the same way as in translating from one written language to another.

Item 29 (Fig. 4.3) is from the same "relations and proportions" test as Item 31, but it was adapted from a recent "National Assessment" examination (Carpenter et al., 1981). To answer item 29 correctly, the student's primary task is to perform a (computational) transformation within the domain of written symbols.

We have found it useful to sort out "between-system mappings" (i.e., translations) from "within-system operations" (i.e., transformations) even though

Item 31. What picture shows ⅓ shaded?

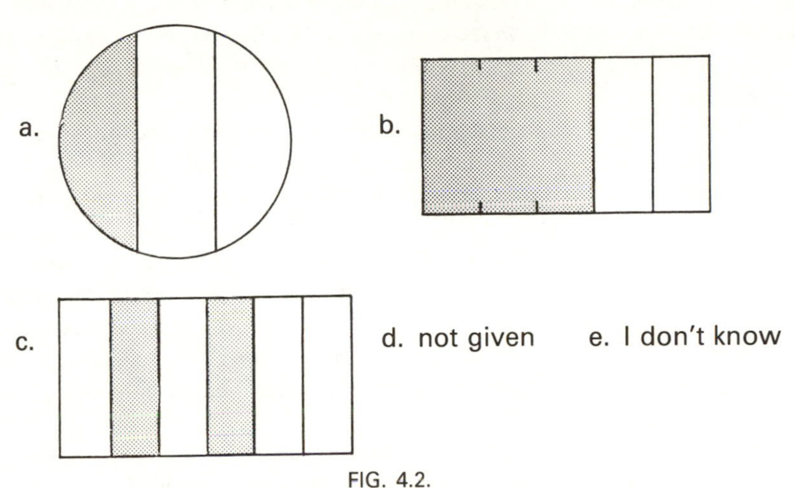

FIG. 4.2.

transformations and translations tend to be interdependent in reality. For example, RN/PR research suggests that students' solutions to item 29 (preceding) typically involve the use of spoken language (together with accompanying translations and transformations) in addition to pure written symbol manipulations (i.e. transformations). On the other hand, our studies also show that repeated drill on problems like 29 does not necessarily provide needed instruction related to underlying translations. For example, consider the following results.

Educators familiar with results from recent "National Assessments" (Carpenter et al., 1981) may not be surprised that our students' success rates for item 29 were only 11% for 4th graders, 13% for 5th graders, 30% for 6th graders, 29% for 7th graders, and 51% for 8th graders. Such performances by American students led to "Nation at Risk" reports from a number of federal agencies and professional organizations. However, success rates on the seemingly simpler

Item 29. The ratio of boys to girls in a class is 3 to 8. How many girls were in the class if there were 9 boys?

a. 17 *b.* 14 *c.* 24 *d.* not given *e.* I don't know

Correct by grade: 4) 11%, 5) 13%, 6) 29.7%, 7) 29%, 8) 51%

Answer Option Selected: *a*) 5.53% *b*) 16.9% *c*) 26.7% *d*) 9.3% *e*) 19.4% *na*) 1.91%

FIG. 4.3.

item 31 were even lower: 4% for 4th graders, 8% for 5th graders, 19% for 6th graders, 21% for 7th graders, and 24% for 8th graders. On the translation item 31, only 1 in 4 students answered correctly! 43% selected answer choice (a); 4% selected (b); 15% selected (c); 34% selected (d); 3% selected (e); and 2% did not give a response.

One major conclusion from our research is apparent from the preceding examples; not only do most fourth- through eight-grade students have seriously deficient understandings in the context of "word problems" and "pencil and paper computations," many have equally deficient understandings about the models and language(s) needed to represent (describe and illustrate) and manipulate these ideas. Furthermore, we have found that these "translation (dis)abilities" are significant factors influencing both mathematical learning and problem-solving performance, and that strengthening or remediating these abilities facilitates the acquisition and use of elementary mathematical ideas (Behr, Lesh, Post, & Wachsmuth, 1985; Post, 1986).

Part of what we mean when we say that a student "understands" an idea like "⅓" is that: (1) he or she can recognize the idea embedded in a variety of qualitatively different representational systems, (2) he or she can flexibly manipulate the idea within given representational systems, and (3) he or she can accurately translate the idea from one system to another. As a student's concept of a given idea evolves, the related underlying transformation/translation networks become more complex; and teachers who are successful at teaching these ideas often do so by reversing this evolutionary process; that is, teachers simplify, concretize, particularize, illustrate, and paraphrase these ideas, and inbed them in familiar situations (i.e., scripts).

To diagnose a student's learning difficulties, or to identify instructional opportunities, teachers can generate a variety of useful kinds of questions by presenting an idea in one representational mode and asking the student to illustrate, describe, or represent the same idea in another mode. Then, if diagnostic questions indicate unusual difficulties with one of the processes in Fig. 4.1, other processes in the diagram can be used to strengthen or bypass it. For example, a child who has difficulty translating from real situations to written symbols may find it helpful to begin by translating from real situations to spoken words and then translate from spoken words to written symbols; or it may be useful to practice the inverse of the troublesome translation, i.e., identifying familiar situations that fit given equations.

Not only are the translation processes depicted in Fig. 4.1 important components of what it means to understand a given idea, they also correspond to some of the most important "modeling" processes that are needed to use these ideas in everyday situations. Essential features of modeling include: (1) simplifying the original situation by ignoring "irrelevant" (or "less relevant") characteristics in order to focus on other "more relevant" factors; (2) establishing a mapping between the original situation and the "model"; (3) investigating the properties of the model in order to generate predictions about the original situation; (4) translating (or mapping) the predictions back into the original situation; and (5) checking to see whether the translated prediction is useful and sensible.

Translation processes are implicit in a variety of techniques commonly used to investigate whether a student ''understands'' a given textbook word problem, e.g., ''Restate it in your own words.'' ''Draw a diagram to illustrate what it's about.'' ''Act it out with real objects.'' ''Describe a similar problem in a familiar situation.'' Or, techniques for improving performance on word problems'' include: (1) using several different kinds of concrete materials to ''act out'' a given problem situation; (2) describing several different kinds of everyday problem situations that are similar to a given prototype concrete model; or (3) writing equations to describe a series of word problems—delaying the actual solutions until the student becomes proficient at this descriptive phase.

Even though representation (or modeling) often tends to be portrayed as involving only a single simple mapping from the modeled situation to the model (or to their underlying concepts, which might be characterized as the skeletons of external structural metaphors), our AMPS, RN, and PR research suggests that the act of representation tends to be *plural, unstable,* and *evolving;* and these three attributes play important roles to make it possible for concepts and representations to evolve during the course of problem-solving sessions. Here are some examples.

In RN and PR research involving concrete/realistic versions of typical textbook word problems, we have found that students seldom work through solutions in a single representational mode (Lesh, Landau, & Hamilton, 1983). Instead, students frequently use several representational systems, in series and/or in parallel, with each depicting only a portion of the given situation. In fact, many realistic problem-solving situations are inherently *multi*modal from the outset. The following two pizza problems illustrate this point.

> Show a 6th grader one-fourth of a real pizza, and then ask, ''If I eat this much pizza, and then one-third of another pizza, how much will I have eaten altogether?''
>
> Show a 6th grader one-third of a real pizza, and then ask, ''If I already ate one-fourth of a pizza, and now eat this much, how much will I have eaten altogether?''

Neither of the preceding problems is a ''symbol–symbol'' or ''word–word'' problem. Instead, the ''givens'' in both problems include a real object (i.e., a piece of pizza), and a spoken word (to represent a past or future situation). Like many realistic problems in which mathematics is used, the situation in these two pizza problems is inherently multimodal. Each of the problems is a ''pizza-word'' problem in which one of the student's difficulties is to translate the two givens into a homogeneous representation mode so that combining is sensible.

Not only may problems of the preceding type occur naturally in a multimodal form but solution paths also often weave back and forth among several representational systems, each of which typically is well suited for representing some parts of the situation but is ill suited for representing others. For example, in the two problems just given, a student may think about the static quantities (e.g., the two pieces of pizza) in a concrete way (perhaps using pictures) but may switch to spoken language (or to written symbols) to carry out the dynamic "combining" actions (Lesh, Landau, & Hamilton, 1983).

Good problem solvers tend to be sufficiently flexible in their use of a variety of relevant representational systems that they instinctively switch to the most convenient representation to emphasize at any given point in the solution process.

Figure 4.4 suggests one way that the act of representation tends to be plural; that is, solutions often are characterized by several partial mappings from parts of the given situation to parts of several (often partly incompatible) representational systems. Each partial mapping represents a "slice" of the problem situation, using only part of the available representational system. It is not a mapping from the whole "given" situation to only a single representational system.

The act of representation also may be plural in a second sense; that is, a student may begin a solution by translating to one representational system and may then map from this system to yet another system, as illustrated in Fig. 4.5.

In fact, for concrete or realistic versions of textbook word problems, the actual solutions our students have tended to use often combine features depicted in both Fig. 4.4 and 4.5 preceding, as well as a third aspect of representational plurality; that is, a given representational system often appears to be related (in a given student's mind) to several distinct clusters of mathematical ideas. An

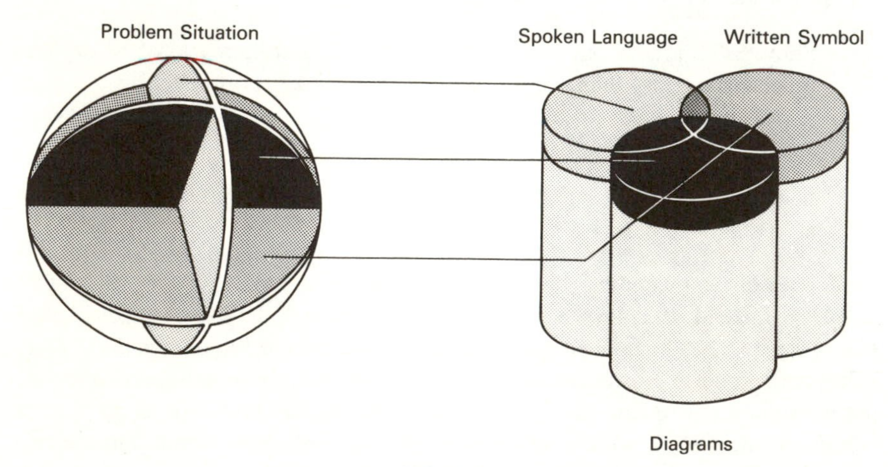

FIG. 4.4.

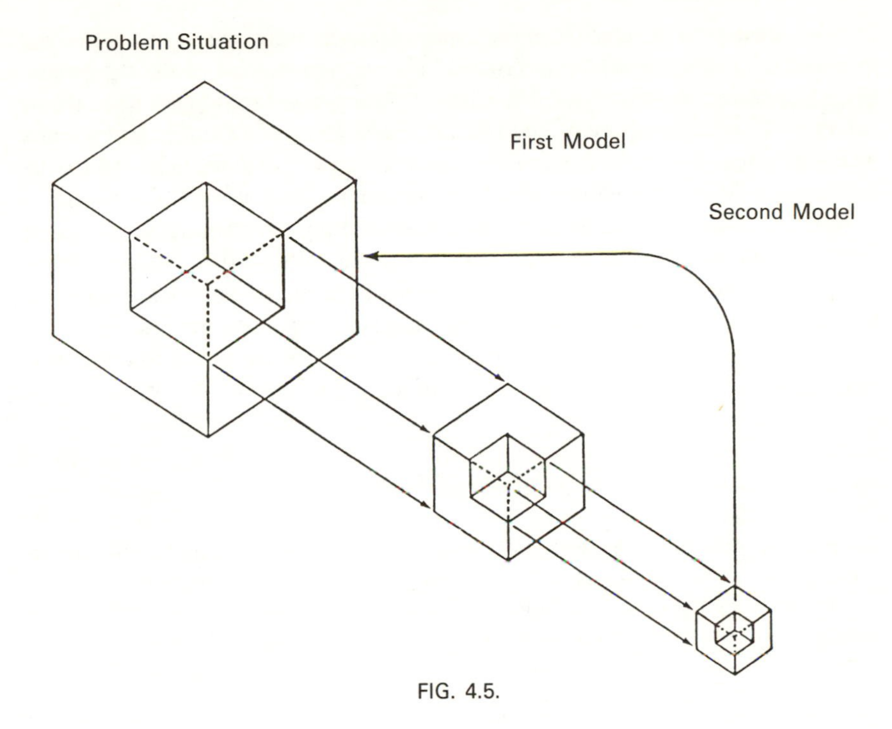

FIG. 4.5.

example to illustrate this point occurred in the AMPS project when several of our students worked on the following ''million dollar'' problem.

> *The Million Dollar Problem:* Imagine that you are watching ''The A Team'' on television. In the first scene, you see a crook running out of a bank carrying a bag over his shoulder, and you are told that he has stolen one million dollars in small bills. Could this really have been the case?

One student who solved this problem began by using sheets of typewriter paper to represent several dollar bills. Then, he used a box of typewriter paper to find how many $1 bills such a box would hold—thinking about how large (i.e., volume) a box would be needed to hold one million $1 bills. Next, however, holding the box of typewriter paper reminded him to think about *weight* rather than *volume*. So, he switched his representation from using a box of typewriter paper to using a book of about the same weight. By lifting a stack of books, he soon concluded that, if each bill was worth no more than $10, then such a bag would be far too large and heavy for a single person to carry.

For the preceding solution, the first representation involved a sheet of paper, which was quickly subsumed into a second representation based on the size of boxes (i.e., volume), which played a role in switching from conceptualizations based on volume to a conceptualization focused on weight. Clearly, the meaning(s) associated with each of these representations were plural in nature; and they *evolved* during the solution process. The unstable nature of the representations was reflected in the fact that when attention was focused on "the whole situation" (or representation), details that previously were noticed frequently were neglected. Or, when attention was focused on one detail (or one interpretation), the student often temporarily lost cognizance of others.

We have addressed the topic of conceptual and representational instability in other RN, PR, and AMPS publications (e.g., Behr, Lesh, Post, & Silver, 1983; Lesh, 1985), so we do not attempt to deal with this rather complex topic here. Instead, we want to stress the inherent plural and evolving nature the act of "representation," because both of these characteristics are linked to the importance of *translations* in mathematical learning and problem solving.

ACKNOWLEDGMENTS

The research of the RN, PR, and AMPS projects was supported in part by the National Science Foundation under grants SED 79–20591, SED 80–17771, and SED 82–20591. Any opinions, findings, and conclusions expressed in this report are those of the authors and do not necessarily reflect the views of the National Science Foundation.

REFERENCES

Behr, M., Lesh, R., Post, T., & Silver, E. (1983). Rational number concepts. In R. Lesh & M. Landau (Eds.), *The acquisition of mathematical concepts and processes*. New York: Academic Press.

Carpenter, T., Corbitt, M., Kepner, H., Lindquist, M., & Reys, R. (1981). *Results from the Second Mathematical Assessment of the National Association of Educational Progress*. Reston, VA: National Council of Teachers of Mathematics.

Lesh, R. (1981). Applied mathematical problem solving. *Education Studies in Mathematics, 12,* 235–264.

Lesh, R., Landau, M., & Hamilton, E. (1983). Conceptual models in applied mathematical problem solving." In R. Lesh, *The acquisition of mathematical concepts and processes*. New York: Academic Press.

Post, T. (1986). *Research based methods for teachers of elementary and junior high mathematics*. Boston: Allyn & Bacon.

Post, T. R., Lesh, R., Behr, M. J., Wachsmuth, I. (1985). Selected results from the rational number project. In L. Streefland (Ed.)., *Proceedings of the Ninth International Conference for Psychology of Mathematics Education:* Vol 1. Individual contributions (pp. 342–351). State University of Utrecht, The Netherlands: International Group for the Psychology of Mathematics Education.

5 Rational Number Relations and Proportions

Richard Lesh
WICAT & Northwestern University

Merlyn Behr
Northern Illinois University

Tom Post
University of Minnesota

This chapter presents examples and some general results from a written test on "relations and proportions" that has been used in our Rational Number (RN) and Proportional Reasoning (PR) projects. The RN and PR projects have included: (1) instructional components in which whole classes of fourth through eighth graders have been carefully observed, tested, and interviewed during "teaching experiments" lasting as long as 20 weeks; and (2) evaluation components in which several thousand second through eighth graders have been tested at a variety of schools in California, Illinois, Minnesota, Pennsylvania, and Utah.

The RN/PR testing program has included three types of tests: pencil-and-paper tests, instruction-mediated tests, and clinical interviews. Each type of test has focused on students' abilities to perform *translations* of a given idea from one representational system to another, or *transformations* within a given representational mode. Our chapter of this book on "representations and translations" gives several reasons why we believe these processes are especially important in mathematics learning and problem solving.

In general, we have been most interested in the "mediating" roles that *spoken language* and *manipulative models* play in problem solving and in the evolution of proportional reasoning and rational number concepts. However, because of the dominant role that pictures, written language and symbols play in textbook-based instruction and testing, we also have investigated these "bookable" representation systems.

The pencil-and-paper phase of the RN testing program consists of items in which *both* the questions and the responses use "bookable" representations. The instruction-mediated testing phase consists of items in which the question presented in spoken language or manipulative models, but the response requires only a "bookable" mode. The clinical interview phase consists of items in which *both* the questions and the responses involve "nonbookable" modes.

The pencil-and-paper testing battery, from which the examples in this chapter were taken, includes a set of three subtests, each focused on a different aspect of rational number understanding: (1) a "basic understandings" test focusing on youngsters' abilities to translate from one representational system to another for a *single fraction* or *ratio;* (2) a "relations and proportions" test, in which the items involve equivalence or ordering relationships among *pairs of fractions* or *ratios;* and (3) an "operations" test with items requiring the subject to "*operate* on" *fraction* or *ratio ideas* embedded in pictures, written mathematics symbols, or written forms of spoken language (e.g., many of the items on this test were either addition or multiplication "word problems" or computation exercises). All the pencil-and-paper items used a multiple choice format.

Details and results from the "basic understandings" and "operations" subtests have been reported in Lesh, Landau, and Hamilton (1983). This chapter focuses on results from the "relations and proportions" test.

GENERAL GOALS OF THE RN/PR TESTING PROGRAM

One goal of the RN/PR testing program has been to provide a data base for comparing and integrating results from past research and to provide baseline student-performance information for future R&D efforts. Whenever possible, we have attempted to use items from other large-scale testing programs (e.g., Carpenter, Coburn, Reys, & Wilson, 1978; Hart, 1980) or from past research or development projects (e.g., Karplus, Pulos, & Stage, 1983; Kieren, 1976; Novillis, 1976; Wagner, 1976). For example, the "relations and proportions" test includes adapted versions of proportional reasoning items from the 1977–78 National Assessment of Educational Progress (1979) and modified versions of "orange juice" items that were used to Noelting and Gagne's (1980) research on proportional reasoning.

Three main goals of this chapter are: (1) to give examples illustrating some of the structural characteristics of students' representational capabilities related to proportional reasoning and rational number understandings; (2) to describe some of the difficulties students frequently experience with representational translations; and (3) to describe several "rules of thumb" about the relative difficulty of several different kinds of representational translations.

The part of the "relations and proportions" test that replicated and extended Noelting's (1980) research aimed at identifying Piagetian "stages" in the development of youngsters' ''proportional reasoning'' concepts. Noelting's research is particularly interesting in this chapter because it deals directly with one of the fundamental themes that runs throughout this book. This issue has to do with the educational importance of analyzing students' operational capabilities for particular mathematical ideas and for specific representational systems that facilitate students' abilities to acquire and use these ideas.

Underlying Philosophy

Piagetian psychologists have been particularly interested in the major conceptual reorganizations that typically occur in children at about age two, at about age six to seven, and during early adolescence. However, mathematics educators tend to be more interested in the transitional phases between (and beyond) these global transition periods. To create learning experiences that will gradually guide students from easier conceptualizations to those that are progressively more complex and abstract, mathematics educators must be able to anticipate the relative difficulties of specific ideas, or alternative conceptualizations of given ideas. They also must be able to anticipate the relative difficulty of different representational systems (or instructional models) that are available to introduce and illustrate these ideas.

Although Piaget was one of the foremost psychologists to emphasize the inherent structural/constructivist nature of mathematical ideas and representational systems, both he and his followers use the term *decalage* to refer to situations in which two ideas (or representations for a particular idea) involve similar underlying structures, but one is significantly easier than the other.

During the past decade, some of the most productive research in mathematics education has focused on tracing the gradual evolution of particular ideas, finding the relative difficulty of distinct conceptualizations of given ideas, and identifying characteristics that influence the difficulty of different models (or representations) of these conceptualizations. In other words, some of the most significant research in mathematics education has focused on finding rules to explain or describe predictable regularities that occur within what Piagetians refer to as decalages.

Results from the Relations and Proportions Test

The following results from the ''relations and proportions'' test were identical to those we have reported earlier based on the ''operations'' and ''basic understandings'' tests (Post, Behr, & Lesh, 1982). Consequently, we simply review them here and give examples to illustrate them using items from the ''relations and proportions'' test. Varying a task on any of a number of dimensions (e.g., number size, relationship of numerator to denominator, perceptual characteristics of figural models, etc.) frequently produces dramatic variations in stu-

Item 24. A map has a scale of 1 inch for every 2 miles. How many miles does 2 inches on the map represent?

e. 3 miles *f.* 2 miles *g.* 4 miles
h. not given *i.* I don't know.

% Correct by grade: 4) 67%, 5) 72%, 6) 82.8%, 7) 87%, 8) 89.4%

Answer Option Selected: *e*) 1.9% f) 6.67% g) 79.7% h) 5.53% i) 5.15% na) 0.95%

FIG. 5.1.

dents' performances. For example, test item 29 (Fig. 5.4) was considerably more difficult than "similar" items like item 24 (Fig. 5.1), which involved "unit fractions or ratios" (like ½ or ⅓), or than items 27, 26, or 60.

Other factors that contribute to item difficulty are related to the "perceptual distractors" that were inherent in certain representations. For example, only about 53% of our eighth graders got item 36 (Fig. 5.6) correct, and only about 62% got item 44 (Fig. 5.7) correct.

Probably the *main* factor that made items 44 and 36 so difficult had to do with the fact that such items presuppose an agreement that must be accepted about the

Item 26. Andre can do 9 chin-ups in 15 seconds. How many chin-ups can he do in 5 seconds?

f. 24 *g.* 3 *h.* 6 *i.* not given *j.* I don't know.

% Correct by grade: 4) 34.9%, 5) 42%, 6) 59.5%, 7) 71%, 8) 85.6%

Answer Option Selected: f) 3.24% g) 58.4% h) 11% i) 17.9% j) 8.02% na) 1.34%

FIG. 5.2.

Item 27. Mrs. Stein's plant uses 2 grams of plant food every 10 days it is in the soil. How long does it take her plant to use 6 grams of plant food?

a. 40 days. *b.* 12 days. *c.* 28 days. *d.* not given. *e.* I don't know.

% Correct by grade: 4 (46.8%), 5 (54%), 6 (55%), 7 (76%), 8 (80.8%).

Answer Option Selected: a) 11.26% b) 8.4% c) 7.44% d) 62.2% e) 8.97% na) 1.72%

FIG. 5.3.

Item 29. The ration of boys to girls in a class is 3 to 8. How many girls were in the class if there were 9 boys?

a. 17 *b.* 14 *c.* 24 *d.* not given *e.* I don't know

% Correct by grade: 4) 11%, 5) 13%, 6) 29.7%, 7) 29%, 8) 51%

Answer Option Selected: a) 5.53% b) 16.9% c) 26.7% d) 29.3% e) 19.4% na) 1.91%

FIG. 5.4.

Item 60. A paint store made a mixture of 10 gallons of yellow paint and 14 gallons of blue paint. How many gallons of blue paint are needed to mix with 5 gallons of yellow paint to produce the same color?

f. 9 *g.* 28 *h.* 7 *i.* not given *j.* I don't know

% Correct by grade: 6) 31.5%, 7) 50%, 8) 50%

Answer Option Selected: f) 15.5% g) 8.57% h) 43.4% i) 14.9% j) 11.7% na) 5.71%

FIG. 5.5.

Item 36. What picture shows ⅔ shaded?

f.

g.

h.

i. not given

j. I don't know

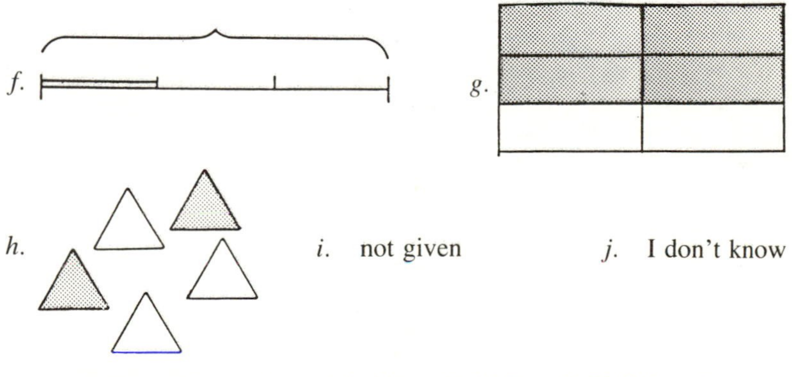

% Correct by grade: 4) 9.2%, 5) 6.1%, 6) 33.3%, 7) 36%, 8) 52.9%

Answer Option Selected: f) 8.78% g) 27.4% h) 12.6% i) 46.9% j) 2.29% na) 1.91%

FIG. 5.6.

Item 44. Which picture shows three-fourths shaded?

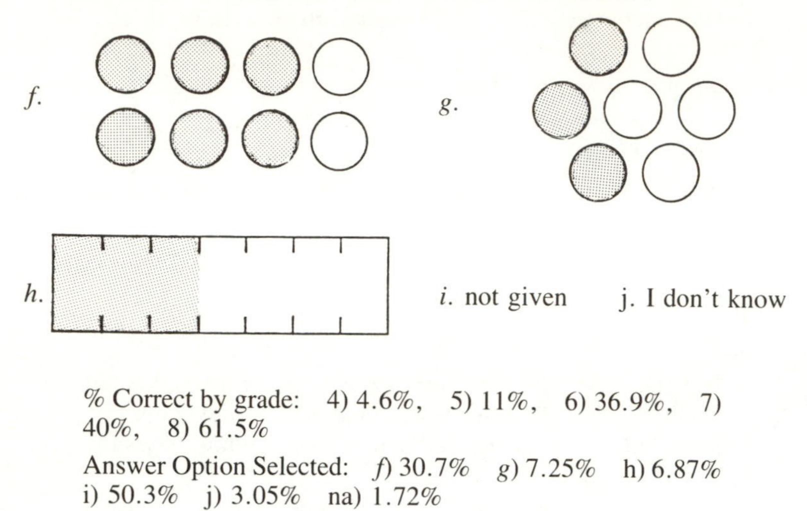

% Correct by grade: 4) 4.6%, 5) 11%, 6) 36.9%, 7) 40%, 8) 61.5%

Answer Option Selected: *f*) 30.7% *g*) 7.25% h) 6.87% i) 50.3% j) 3.05% na) 1.72%

FIG. 5.7.

basis for calling two situations equivalent. For example, to a child who wants to play an electronic arcade game, four quarters may not be "equivalent to" one dollar; similar observations could be made about fractions and ratios in many everyday situations. However, this is not the only factor that seemed to be involved.

In spite of the preceding comments, we should not be too hasty in accepting the notion that students who did not give "correct" answers to items 44 and 36 did so only because they were thinking of a more restricted interpretation of equivalence. For example, many of the students who responded "incorrectly" to items 44 and 36 used the criteria we were expecting in other situations, like items 43 or 37 (see Figs. 5.8–5.9).

To explain the behavior of these students, it is necessary to explain how the criteria they use to judge "equivalence" is influenced by characteristics of the task. Why are they inconsistent? Our research suggests that students who use one criteria in one situation but abandon it in other "similar" situations apparently do so because of the strength of "perceptual distractors" (e.g., "extra partitions") that were built into some diagrams more than others (Behr, Lesh, Post, & Silver, 1983; Post, Wachsmuth, Lesh, & Behr, 1985).

An even more dramatic example of the influence of perceptual distractors occurred in the clinical interview phase of the RN/PR testing program. Students were asked to "Give me one-third of this Hershey chocolate bar." The difficulty of this problem depended significantly on whether the student was given a "plain" Hershey bar, or a Hershey bar "with nuts." The plain bar was signifi-

Item 37. What fraction of the circle below is shaded?

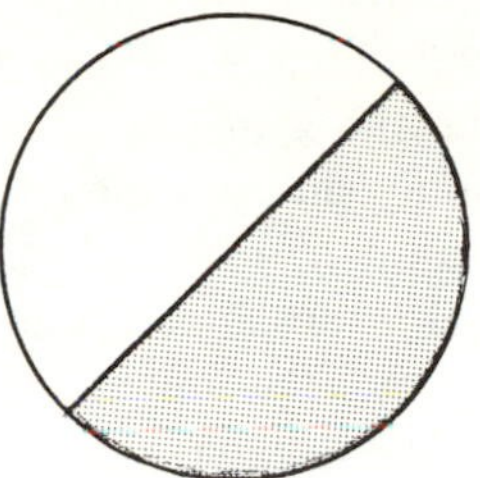

a. 1 *b.* 2 *c.* 5/10 *d.* not given *e.* I don't know

% Correct by grade: 4) 9.2%, 5) 13%, 6) 36%, 7) 48%, 8) 57.7%

Answer Option Selected: a) 10.1% b) 3.24% c) 32.6% d) 43.3% e) 9.35% na) 1.34%

FIG. 5.8.

cantly more difficult, apparently because of the partitions on the plain bar that were not on the bar with nuts.

The fact that students' rational number thinking is often so strongly influenced by "perceptual distractors" suggests that their underlying conceptualizations of many rational number ideas are quite unstable (Behr, Wachsmuth, Post, & Lesh, 1984).

Because each item on our three pencil-and-paper tests can be characterized by one of the following seven translations (see the chapter of this book on "representations and translations" for more information about these seven translation types), it is possible to think of the RN pencil-and-paper testing battery as

Item 43. What fraction is shaded in the picture below?

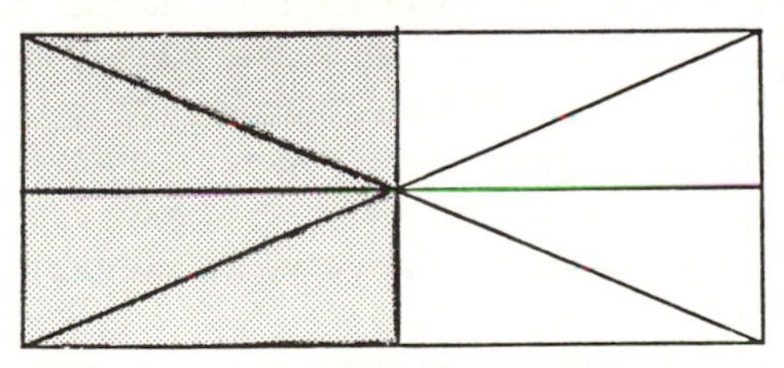

a. 4/4 *b.* 1/2 *c.* 1/1 *d.* not given *e.* I don't know

% Correct by grade: 4) 20%, 5) 38%, 6) 59.5%, 7) 78%, 8) 86.5%

Answer Option Selected: a) 9.92% b) 56.1% c) 2.67% d. 28.4% e) 1.34% na) 1.53%

FIG. 5.9.

consisting of seven "translation subtests"—one for each type of translation: (a) symbols to written language, (b) written language to symbols, (c) pictures to pictures, (d) written language to pictures, (e) pictures to written language, (f) symbols to pictures, (g) pictures to symbols.

If the pencil-and-paper testing battery is thought about in the preceding way, then the order of increasing difficulty of the "translation subtests" was as just listed, with "picture to symbol" translations being most difficult. In general, if other factors are held constant: (a) Translations to pictures is easier than translations from pictures; (b) translations involving written language (e.g., three-fourths) are easier than translations involving written symbols (e.g., ¾); and (c) the easiest translations are those that only require a student to "read" a fraction or ratio in two different written forms. The difficulty (measured in % correct) of an item did not necessarily decrease as grade level increased. For example, on item 29 (Fig. 5.4), seventh-grade performance was no better than sixth-grade performance. In fact, on all three of our pencil-and-paper translation tests, the greatest proportion of performance decreases occurred between sixth and seventh grades, whereas the greatest proportion of "large performance increases" occurred between the fifth and sixth grades. The rational number understandings of seventh graders seemed to be particularly unstable. Why? What explanations might account for such phenomena?

In Lesh, Landau, and Hamilton (1983), one-to-one researcher-to-child interview results on problems similar to item 29 (Fig. 5.4) suggested that seventh graders are shifting to a new solution mode (i.e., more algebraic/symbolic) than those that are more typical among sixth graders and that this "shift" brings with it some new sources of error. Seventh graders may be more likely to use formal/symbolic solution procedures rather than more concrete or intuitive approaches—so they may experience difficulties on items that do not readily fit familiar equations. For example, notice that the statement of item 29 is not of the form "3 is to 8 as 9 is to x." So, students who attempted to use an a/b = c/d style of equation may have been attempting to find a number such that "3 is to 8 as x is to 9 (i.e., 3/8 = x/9). If this explanation is valid, however, we are again faced with the problem of explaining why other similar similar problems were relatively easy. Again, our follow-up interviews suggest that inconsistent behavior seems to be the result of facilitating or distracting factors that are "built into" certain diagrams or representations. More examples to illustrate this claim are given next.

Every item on the three pencil-and-paper tests was classified according to the following five dimensions: (1) representational mode, or translation type, (2) rational number subconstruct (i.e., ratio versus fraction), (3) figural attributes (e.g., discrete versus continuous elements, contiguous parts versus noncontiguous parts, etc.), (4) "Piagetian" level ala Noelting (see next section for a brief explanation of this dimension), and (5) format, e.g., whether the item was: (a) an equation translation problems, like item 34 (Fig. 5.10), (b) a fraction or

Item 34. What fraction sentence says the same thing as this picture?

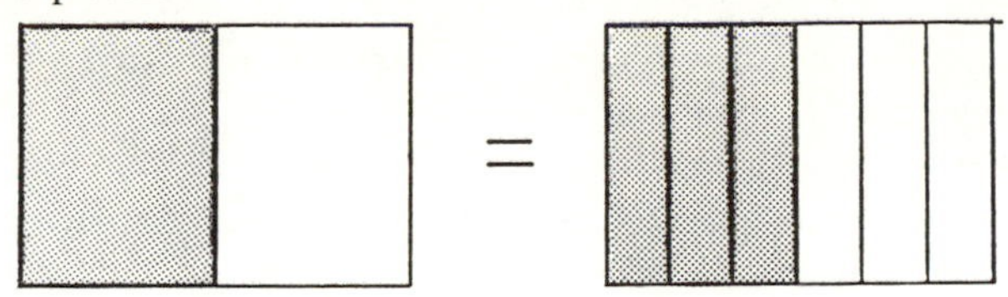

f. $1/2 = 3/6$ *g.* $1/1 = 3/3$ *h.* $2/1 = 6/3$ *i.* not given *j.* I don't know

% Correct by grade: 4) 70.6%, 5) 81%, 6) 87.4%, 7) 89%, 8) 96.2%

Answer Option Selected: f) 84.7% g) 4.2% h) 3.82% i) 2.1% j) 3.44% na) 1.72%

FIG. 5.10.

ratio translation problem, like items 36 or 44 (Figs. 5.6–5.7), (c) a proportion word problem, like items 24 or 29 (Figs. 5.1–5.4), or (d) a comparison problem, like the ones that are described in the next section of this chapter about Noelting and Gagne's Piagetian Levels.

Distinctions between fractions and ratios, and between discrete and continuous elements are illustrated in item 44. Answer choice *g* is an example of a "discrete" model, whereas *h* is a "continuous" model. This discrete-versus-continuous distinction has been a task variable influencing item difficulty on all three pencil-and-paper tests and in a number of other past research studies (e.g., Behr et al., 1983; Novillis, 1976).

Fraction-ratio confusions also appeared to be sources of difficulty for many items. For example, among the incorrect responses on item 44, 14% selected either answer choice *g* or *h*—both of which involve *ratios* of 3-to-4 between shaded and unshaded parts.

Clear performance differences appeared to be associated with all the aforementioned "task variable dimensions." However, the effect of these variables was not necessarily additive; that is, the impact of changing a single variable in a pair of items may either be much greater than, or possibly even opposite to, a similar change for another pair of items. It seems likely that any explanation that attempts to predict the difficulty levels of tasks like those on the "relations and proportions" test cannot be "one dimensional" and surely must take into account the facilitating and distracting characteristics of various representational systems.

The next section of this chapter focuses on one scheme, based on the work of Piaget, that attempts to explain the difficulty of one subset of the items on the "relations and proportions" test, i.e., those dealing with proportional reasoning.

Comparison Problems and Noelting's Piagetian Levels

Thirty-six items on the "relations and proportions" test involved either comparing the "orange flavor" of two orange juice mixes, or comparing the size of fractions represented in pictures or symbols.

Item 21 (next) is an example of a question about making an orange drink by mixing orange concentrate and water. The shaded squares stood for orange concentrate, and the white squares stood for water. The student was to select the true statement following each picture.

Item 21 is an example of a ratio comparison with continuous quantities, and item 65 (Fig. 5.13) is a fraction comparison.

Noelting and Gagne (1980) devised a Piagetian-based scheme to predict the difficulty levels of proportional reasoning and fraction comparison tasks such as those just shown. Their results showed that this scheme was stable with respect to the three types of items, namely orange juice comparisons (e.g., item 21, given earlier), fraction comparison depicted with symbols, and fraction comparisons represented with pictures (e.g., item 65, following).

We used the following scheme, which is mathematically equivalent to Noelting and Gagne's, and which applies to any of the ratio or fraction comparison items on the "relations and proportions" test:

Level 1. two unequal rationals, varying in either the first component or the second component, but not both (e.g., one-third versus two-thirds, or three-fifths versus three-sevenths.)

Level 2A. two equal unit rationals (e.g., one-fourth versus two-eighths).

Item 21. You are making orange juice by mixing orange concentrate and water. The shaded squares stand for orange concentrate and the white squares stand for water. Select the true statement following the pictures.

A B

a. Mixture A has a stronger orange flavor than Mixture B.
b. Mixture B has a stronger orange flavor than Mixture A.
c. Mixture A and Mixture B have the same orange flavor.
d. I don't know which mixture has the stronger orange flavor.

% Correct by grade: 4) 7.3%, 5) 12%, 6) 35%, 7) 34%, 8) 50%

Answer Option Selected: a) 17.3% b) 52% c) 27.6% d) 2.1% na) 0.76%

FIG. 5.11.

Item 65. You are comparing the fractions in pictures A and B. Select the true statement following the pictures.

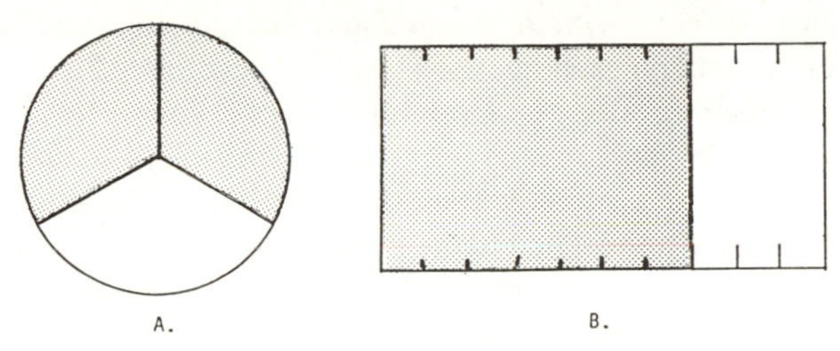

a. A shows a greater fraction shaded than B.
b. B shows a greater fraction shaded than A.
c. A and B show the same fraction shaded.
d. I don't know which picture shows the greater fraction.

% Correct by grade: 7) 45%, 8) 54.8%,

Answer Option Selected: a) 26.4% b) 50% c) 17.1% d) 3.43% na) 2.94%

FIG. 5.12.

Level 2B. two equal nonunit rationals (e.g., three-fifths versus six-tenths).

Level 3A. two unequal rationals with at least one pair of "between" or "within" components divisible (e.g., two-thirds versus four-fifths with divisibility "between" the two and the four; two-fifths versus three-ninths—with the divisibility "within" the three-ninths).

Level 3B. two unequal rationals, with no between or within component divisibility (two-thirds versus three-fifths, for example).

Results

Table 5.1 assigns a "Noelting level" to all the rational number comparison items in the relations and proportions test, with these items partitioned into four categories: "orange juice" comparisons, "picture–picture" comparisons, symbol–symbol comparisons, and "other" (i.e., comparisons that are not part of the Noelting format, but which can be classified using his scheme.)

Within each category, on Table 5.1, the items are ordered according to difficulty, so that difficulty-comparisons are easy to make, both within categories, and across categories. For example, item 21 (Fig. 5.11), which is an "orange juice" comparison identified with Noelting level 2B, is positioned at point 63 on the "orange juice" line—because it was the 63rd most difficult item on the relations and proportions test. Similarly, item 65 (Fig. 5.12), which is a picture–picture comparison, is identified with a "3B" at the point 59 on the picture–picture line—because it was the 59th most difficult item on the test.

In a crude way then, Table 5.1 makes it possible to compare the difficulty of any of the comparison items, or any two items on the test.[1] In particular, any "reversals" from Noelting's predictions are apparent. If our students' performances matched perfectly with the predictions of the Noelting and Gagne model, then, from left to right on each line, one would expect to read the classifications in consecutive groupings (i.e., all the level 1 items followed by all the level 2A items, then all the 2B items, etc).

Table 5.1 shows that, within the first three categories, the order of difficulty of the items in general correspond to the sequences of difficulty hypothesized by Noelting and Gagne. Indeed, even though a number of reversals can be found on each of the first three number categories, the hypothesized sequences were statistically valid as the first three categories were merged. However, in the "other" category, which included comparisons not given in the Noelting/Gagne format, the hypothesized ordering did not come close to reaching statistical significance; and if these data were merged with those of the other three categories, then .05 statistical significance was not attained.

Our results, then, are not inconsistent with Noelting and Gagne's finding of *statistically* significant sequences based on the "numerical characteristics," as long as attention is restricted to the kinds of tasks used in the earlier studies. However, if a slightly larger class of items is included, where more factors related to particular representations are involved, the hypothesized sequences break down. This suggests that the difficulty-causing dimension identified by Noelting and Gagne probably plays a significant role, but that properties of the representational system also may need to be considered.

Post hoc analyses and follow-up clinical interviews suggested that exceptions to Noelting and Gagne's hierarchy were, in general, explainable on the basis of representational variations. Two examples are given next.

Item 23 (Fig. 5.13), at stage 3b, proved to be easier than all but one of the stage 2B, stage 3A, and other stage 3B orange juice items. As a "2/3 versus 5/8" comparison, the Noelting/Gagne hierarchy predicts that item 23 should be much more difficult than (for example) item 21 (Fig. 5.11), which is a "3/1 versus 6/2" (level 2A) comparison. However, the arrangement of objects in item 23 facilitates one-to-one correspondences between orange juice and water, enabling more intuitive (and correct) judgments than the arrangements for the mathematically less complicated item 21.

For the picture fraction comparisons, item 33 (Fig. 5.14), involves "½" and consequently was put into a special "easy" category according to Noelting's scheme. This is because past research has shown "halfness" to be a cognitively

[1]More sophisticated ordering techniques would have to take into account, for example, that the order of difficulty of the items was not consistent across grade level. We elected to use Table 5.1 because of its intuitive appeal in illustrating the general nature of the "degree of difficulty" relationships among items. The comparisons in Table 5.1 are strictly ordinal and did not enter into any statistical analyses that were applied to the data.

TABLE 5.1
Relative Difficulties of Items Characterized By Noelting Levels

Orange juice comparisons:

1	1	2B	3B	3A	3A	2A	2B	3A	3B	2B	3A	Noelting Level
19	46	41	54	50	63	18	44	64	62	49	25	Item Number

Picture-picture comparisons:

1	1	2A	2B	2B	1	H*	3B	3B	3A	H*	Noelting Level
9	3	10	15	20	28	13	68	55	60	69	Item Number

Symbol-symbol comparisons:

1	1	2B	2B		2A		3A		3B	3B	Noelting Level
				2A		3A		3A			
12	61	26	52		40		67		59	37	Item Number
				17		14		66			

Comparisons not given in Noelting format:

2A	H*		H*		2B		2B		2A		H*	2B	2B	2B	3B		2B		2A	H*	2B	2B	2B	2A		2B			2B	Noelting Level
		2A		H*		2B		2B		2B						H*		2A							2B		2A	2A		
58	34		39		45		27		11		30	5	4	23	68		56		33	7	65	8	70	57		36			31	Item Level
		38		24		6		42		43						53		22							21		29	48		

*''H'' refers to comparisons involving ''one-half'' (stage 2A in Noelting scheme)

primitive concept; youngsters typically show earlier proficiency for tasks involving one-half than for other fraction items (Kieren, 1976). Yet, item 33 was much more difficult than item 10 (Fig. 5.15), which Noelting's scheme would classify at stage 2A. Both items 10 and 33 (Fig. 5.14) involved nonadjacency of parts; however in item 10, the one-third portion in picture A could be (mentally) partitioned to form the three one-ninth portions in picture B; that is, the three one-ninth portions can be (mentally) "pushed together" to form a piece congruent to the one-third portion in picture A. In item 33, such a partitioning was not possible. The one-half portion in part A cannot be partitioned to form the three one-sixth portions in part B; "pushing" the pieces together in picture B would produce a one-half portion cut horizontally, not vertically, as in picture A. Therefore, successful completion of item 33 requires mapping both picture A and picture B to the idea "one-half," and then making a judgment based on transitivity ("A is one-half and B is one-half, therefore A and B are equivalent."). In item 10, it is possible to map each of pictures A and B to the idea "one-third," and then to use transitivity reasoning to complete the comparison, but it is natural (and much easier) to determine equivalence by visual judgment.

In each such case where the Noelting sequence was violated, plausible explanations based on nuances in the item representations seem to be apparent.

Conclusions

Results from the RN/PR "relations and proportions" test have been used to illustrate a number of characteristics of students' representational capabilities.

Item 23. You are making orange juice by mixing orange concentrate and water. The shaded squares stand for orange concentrate and the white squares stand for water. Select the true statement following the picture.

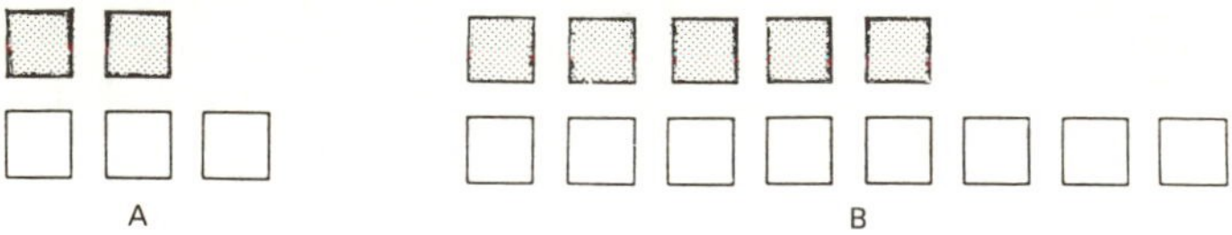

A B

a. Mixture A has a stronger orange flavor than Mixture B.
b. Mixture B has a stronger orange flavor than Mixture A.
c. Mixture A and Mixture B have the same orange flavor.
d. I don't know which mixture has the stronger orange flavor.

% Correct by grade: 4) 33%, 5) 44%, 6) 45%, 7) 48%, 8) 51%

Answer Option Selected: a. 44.2% b. 43.1% c. 6.87% d. 4.58% na. 1.15%

FIG. 5.13.

Item 33. What fraction is shaded in the picture below?

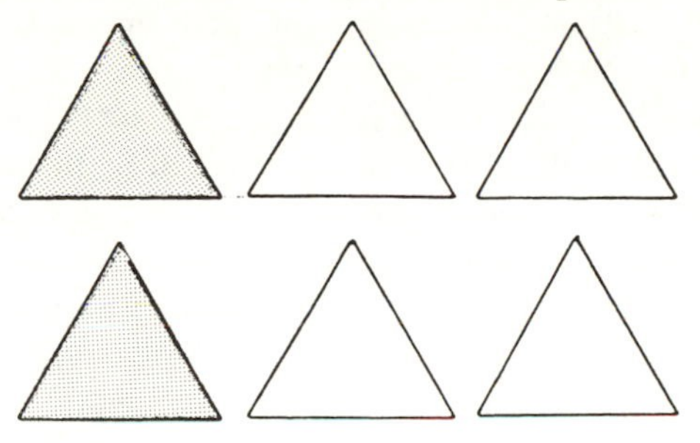

a. one-third *b.* four-halves *c.* six-halves *d.* not given *e.* I don't know

% Correct by grade: 4) 22%, 5) 14%, 6) 31%, 7) 50%, 8) 68.3%

Answer Option Selected: a) 37% b) 0.95% c) 4.96% d) 52.8% e) 2.67% na) 1.53%

FIG. 5.14.

Item 10. You are comparing the fractions in pictures A and B. Select the true statement following the pictures.

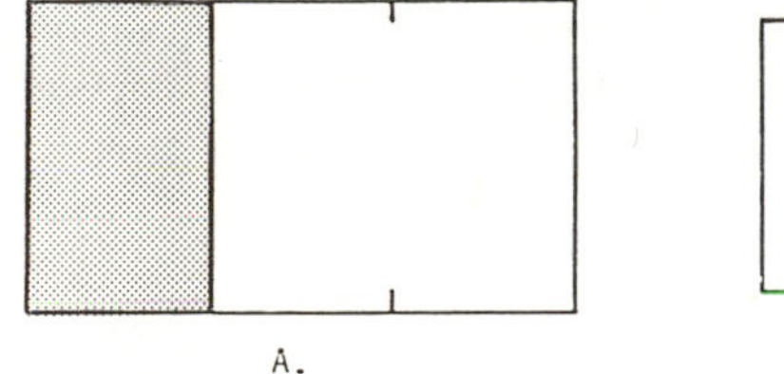

a. A shows a greater fraction shaded than B.
b. B shows a greater fraction shaded than A.
c. A and B show the same fraction shaded.
d. I don't know which picture shows the greater fraction.

% Correct by grade: 4) 51.4%, 5) 60%, 6) 65.8%

Answer Option Selected: e) 22.8% f) 13.7% g) 59% h) 3.75% na) 0.63%

FIG. 5.15.

We also have identified a number of relationships among students' rational number understandings and their abilities to perform translations among and transformations within various representational systems. Whereas "underlying mathematical structure" and "concreteness of task" are two important variables contributing to item difficulty, the characteristics of particular representational systems in which these mathematical structures are embedded also are important.

Mathematics educators are familiar with "number–numeral" distinctions and similar distinctions related to the idea that mathematics ideas are not "in" things but rather are "pure structures." Nonetheless, representational systems can be viewed as being transparent or opaque. A transparent representation would have no more nor less meaning than the idea(s) or structure(s) they represent. An opaque representation would emphasize some aspects of the idea(s) or structure(s), and de-emphasize others; they would have some properties beyond those of the idea(s) and structure(s) that are embedded in them, and they would not have some properties that the underlying idea(s) and structure(s) *do* have. In our research, the opaque nature of representations has been salient; this is one reason why, in real or concrete problem-solving situations, we have found that students often make simultaneous use of several qualitatively different representational systems (e.g., a picture, spoken language. and written symbols), and that the salience of one system over others frequently varies from stage to stage in solution attempts. Capitalizing on the strengths of a given representational system, and minimizing its weaknesses, are important components of "understanding" for a given mathematical idea. Representational translations and transformations are important both to the acquisition and use of mathematical ideas.

If translation abilities are such obvious components of mathematics understanding and problem solving, why are they so often omitted from instruction and testing? One reason is that many translation types are not easily "bookable"; other reasons stem from the fact that so many research questions remain unresolved concerning the exact roles that translations play in the acquisition and use of mathematical ideas, and about the instructional outcomes that can be expected if they are taught effectively (Behr et al., 1983). The following popular misconceptions also are relevant.

Misconception. Translations often are assumed to be easy. *Our research shows* that concrete problems often produce *lower success* rates than comparable "word problems" or written symbolic problems. Lesh, Landau, and Hamilton (1983) includes examples of word problems that become *more difficult* when additional information is given in the form of concrete materials (which one might naively suppose should have made the problems more meaningful, and perhaps easier).

Misconception. If a student can correctly answer a given type of word problem, he or she surely must be able to solve similar problems in everyday situations. This is because word problems are "abstract" versions of real situations. *Our research shows* that purportedly realistic word problems often differ significantly from their real-world analogues with respect to processes most often used in solutions, the types of errors that occur most frequently, and difficulty (Lesh et al., 1983). It is too simplistic to think of word problems as lying on a continuum between abstract ideas and real situations; a word problem may involve understandings that its real analogue does not, and vice versa.

More than 40 weeks of student activities have been developed in the RN and PR projects, based on the same theoretical framework as that used to generate items for the testing program. Consequently, profiles of abilities and understandings that can be generated from the testing program can be used to select appropriate learning experiences for individual children; or, the testing materials can be used to measure effects of theory-based instructional treatments. These sorts of possibilities are beginning to be explored using microcomputer-based instructional materials currently being developed at the World Institute for Computer Assisted Teaching (WICAT).

ACKNOWLEDGMENT

The research of the RN and PR projects was supported in part by the National Science Foundation under grants SED 79–20591 and SED 82–20591. Any opinions, findings, and conclusions expressed in this report are those of the authors and do not necessarily reflect the views of the National Science Foundation.

REFERENCES

Behr, M., Lesh, R., Post, T., & Silver, E. (1983). Rational number concepts. In R. Lesh & M. Landau (Eds.), *Acquisition of mathematics concepts and processes*. New York: Academic Press.

Behr, M., Wachsmuth, I., Post, T., & Lesh, R. (1984). Order and equivalence relations: A clinical teaching experiment. *Journal for Research in Mathematics Education, 15* (5), 323–341.

Carpenter, T., Coburn, T. G., Reys, R. E., & Wilson, J. W. (1978). *Results from the first mathematics assessment of the National Assessment of Educational Progress*. Reston, VA: The National Council of Teachers of Mathematics.

Hart, K. (1980). *Secondary School Childrens' Understanding of Mathematics* (Research Monograph). Chelsea College, England.

Karplus, R., Pulos, S., & Stage, E. (1983). Proportional reasoning of early adolescents. In R. Lesh & M. Landau (Eds.), *Acquisition of mathematical concepts and processes*. New York: Academic Press.

Kieren, T. (1976). On the mathematical, cognitive, and instructional foundations of rational numbers. In R. Lesh (Ed.), *Number and measurement: Papers from a research workshop*. Columbus, OH: ERIC/SMEAC.

Lesh, R., & Akerstrom, M. (1981). Applied problem solving: A priority focus for mathematics education research. In P. Lester (Ed.), *Mathematical problem solving*. Philadelphia: Franklin Institute Press.

Lesh, R., Landau, M., & Hamilton, E. (1983). Conceptual models in applied mathematical problem solving. In R. Lesh & M. Landau (Eds.), *Acquisition of mathematics concepts and processes.* New York: Academic Press.

Noelting, K. (1980). The development of proportional reasoning and the ratio concept. Part I: Determination of stages. *Educational studies in mathematics* (Vol. II, pp. 217–153).

Noelting, K., & Gagne, R. (1980). The development of proportional reasoning in four different contexts. *Proceedings of the Fourth International Conference for the Psychology of Mathematics Education.* Berkeley, CA.

Novillis, C. (1976). An analysis of the fraction concept into a hierarchy of selected subconstructs and the testing of the hierarchical depencencies. *Journal for Research in Mathematics Education.*

Post, T., Behr, M., & Lesh, R. (1982). Issues in the teaching and learning of rational number concepts. *1982 Yearbook of the National Council of Teachers of Mathematics.* Reston, VA: National Council of Teachers of Mathematics.

Post, T., Wachsmuth, I., Lesh, R., & Behr, M. (1985). Order and equivalence of rational numbers: A cognitive analysis. *Journal for Research in Mathematics Education, 16* (1), 18–36.

The Second Assessment of Mathematics, 1977–1978; Released Exercise Set. (1979, Mag). The National Assessment of Educational Progress, Suite 700, 1860 Linoln Street, Denver, CO.

Wagner, H. (1976). An analysis of the fraction concept. *Dissertation Abstracts International, 36,* 7919A–7920A (University Microfilms No. 76–13, 111).

6 Levels of Language in Mathematical Problem Solving[1]

Gerald A. Goldin
Rutgers University

INTRODUCTION

The purpose of this chapter is to elaborate on a model for mathematical problem solving based on four levels of language processing (Goldin, 1982). The intent of the model is to provide a unified framework within which competency in problem solving can be discussed, teaching objectives can be formulated, and observed problem-solving processes can be simulated. The model's origins lie in the conclusions of an exhaustive study of task variables in mathematical problem solving (Goldin & McClintock, 1979). Taking the viewpoint that the task functions as a measuring instrument in research, evoking the problem-solving processes that are to be recorded and studied, it becomes essential to understand how characteristics of the task affect these processes. The following main categories of task variables are observed to have such effects: (1) syntax variables, describing the grammatical structure and complexity of the problem statement; (2) content and context variables, describing, respectively, the mathematical and nonmathematical semantic content of the problem; (3) structure variables, describing the characteristics of formal problem state-space representations and algorithmic procedures; and (4) heuristic behavior variables, describing the heuristic processes that are logically applicable to the problem and the consequences of applying them. Much is known about *which* changes within these categories of task characteristics can affect problem-solving outcomes, and *what* effects are observed to occur. The present model seeks to address directly the

[1]Reprinted with minor editorial revisions from Bergeron, J. C. and Herscovics, N., eds. (1983). *Procs. of the Fifth Ann. Mtg. of PME-NA* (N. Amer. Chapter of the Int'l. Gp. for the Psych. of Math. Educ.), Vol. 2, pp. 112–120. Montreal: Concordia Univ. Dept. of Mathematics.

question of *how* each family of task variables produces its observed effects. Consequently, there is a close association between the levels of language processing comprising the model, and the identified groupings of task variables.

The model is strongly motivated by information-processing theories. Proceeding on the assumption that a meaningful simulation of the human being as an information processor should be based on higher level languages, rather than the "machine language" level of neural nets, it attempts a preliminary description of those higher level languages that might be needed. The term *language* is understood here in the broad sense of a system for processing symbol-configurations, not in its everyday meaning of a system for verbal intercourse. The model's framework is constructed with the eventual goal of accounting for a large portion of the experimental findings about problem solving in the psychological and educational literature, e.g., the roles of strategies and heuristic processes, "unconscious" processing, spatial visualization, pattern recognition, and "insight." With these considerations in mind, then, it is proposed that at least the following four higher-level language systems are entailed: (1) a system for verbal and syntactic processing of "natural" language, accounting for effects of task syntax variables; (2) a system for imagistic (nonverbal) processing, accounting for effects of task content and context variables; (3) a system for generating moves within formal notational languages, accounting for effects of task structure variables; and (4) a system for planning and executive control, accounting for effects of task heuristic behavior variables. Output from any one of these systems, appropriately transformed, should be able to serve as input for any of the others. Figure 6.1 summarizes some of the relationships that exist among the language systems in this model.

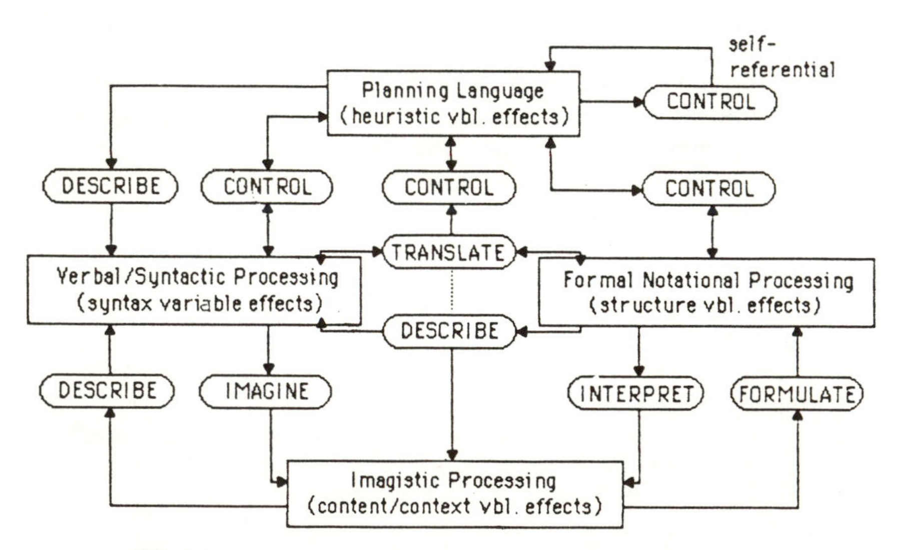

FIG. 6.1. A Model for Competency in Mathematical Problem Solving.

Here I discuss briefly each language system. Before proceeding, it should be noted that many features of human information processing are omitted from this description—for example, the model does not address questions of how information is stored in memory and retrieved, nor does it include the role of "affect" in problem solving. These are possible elaborations of the theory in the future.

VERBAL AND SYNTACTIC PROCESSING

Problem solvers can process a problem's verbal statement in ordinary language such as French or English. At least some of this processing can be described on a purely verbal and syntactic level. Solvers interpret the grammar of sentences and associate words and phrases with other words and phrases. Question sentences are distinguished from declarative statements, and information is sorted to some degree by virtue of syntax into that which is given and that which is desired. One can envision a computer programmed with dictionary information (definitions, synonyms, attribute lists, word associations, etc.) and grammatical rules (procedures for parsing sentences, obtaining referents of pronouns, etc.) that would permit considerable processing on this purely verbal level. Such processing could account for observed effects of manipulation of syntax variables, although up to now such effects have been studied mainly in linear regression models that often include effects due to other variables as well.

Let us consider inputs to and outputs from this system. Of course input can be a verbal statement communicated in print or orally. But solvers also have the capability to describe a nonverbal configuration in words. Here I call this process DESCRIBE, and envision it as a kind of subroutine whose input can be an imagistic, formal, or planning level configuration, and whose output is a verbal statement serving as input to the verbal/syntactic processing level. The configurations in the domain of DESCRIBE can be regarded as modeling those of which one is "conscious" during problem solving. One possible output of verbal/syntactic processing might be a formal notational configuration. That is, a learned procedure I call TRANSLATE can accept as input a verbal statement and output on arithmetic statement or algebraic equation. Many teachers seem to regard this as the main objective in solving "word problems," and the well-known STUDENT program of Bobrow (1968) models this idea very well. However, this process may well be the antithesis of what we mean by "insightful" problem solving. The pitfalls of such direct translation from the problem statement into mathematical notation include children's and teachers' frequent resort to "key words" as shortcuts for selecting arithmetic operations, without comprehension of the problem situation. An alternative procedure is to transform a verbal statement into an "imagistic" configuration. That is, a process I have labeled IMAGINE can accept verbal/syntactic configurations as input and can output nonverbal representations that are subject to imagistic processing.

IMAGISTIC PROCESSING

Sometimes almost involuntarily, problem solvers "imagine" a situation described by a verbal problem statement—"visualize" it, "feel" it, etc. The word "imagistic" that I have selected to stand for such nonverbal encoded configurations is not intended to suggest purely visual images, but to have the connotations of the word "imagine." There seem to be strong reasons for calling this level a language system, and including it as such in the model, though this modifies a perspective I took earlier (Goldin, 1982). There is abundant evidence in the psychological literature of the importance of factors relating to "spatial visualization" in mathematical ability. Problem solvers describe themselves as, for example, "mentally rotating" a figure. In the realm of imagistic processing one may also include pattern recognition, matching nonverbal sensory inputs with previously encoded information, one of the most powerful components of successful problem solving. Thus it appears that problem solvers process crucial information in ways that cannot be understood as purely verbal. On the other hand, such processing probably should not be relegated to a strictly "neural" level. A person may not be able to say "how" he or she "sees" a triangle when encountering the word, yet may be able to describe in detail the consequences of transforming the triangle in a variety of ways. The initial step of visualizing the triangle is undoubtedly very complicated on the neural level but may be usefully regarded as a single step in a higher level imagistic language.

One reason that imagistic processing is difficult to understand from the information-processing viewpoint is that we do not know how imagistic configurations are encoded. It is not easy to describe the "visual image" that a person claims to have as a sequence of symbols, nor is it valid to assume that the brain functions as a miniature movie screen on which images are projected and manipulated. Pattern-recognition, furthermore, is a very elusive notion on the level of explaining *how* one manages to extract essential features from one configuration and identify them as "the same" as essential features from another. These are central questions for study.

Imagistic processing is conceptualized as that system whose operation accounts for the effects of manipulating task content and context variables. It is an interesting conjecture that "genius" in mathematics or physics has to do mainly with the imagistic level. It seems unlikely that the unusually great mathematician achieves that greatness merely by superior encoding of verbal problem statements in formal symbols, or by more efficient processing of formal notations, or by much better and more sophisticated heuristic plans. It is more plausible that the innovator has somehow succeeded in constructing a superior imagistic (nonverbal) representation, within which mathematical relationships inaccessible to others can be "seen." The challenge of mathematics for such a person is to develop appropriate notations within which such relationships can be precisely described and axiomatized, so that theorems can be proven.

FORMAL NOTATIONAL PROCESSING

Notations available for mathematical problem solving are highly structured formal systems. Learning their use probably comprises more than 95% of present-day school mathematics. Once a problem has been translated into such a notational language, purely formal steps can solve it, though it may still be necessary to employ sophisticated search strategies (state-space search algorithms, etc.) in order to decide effectively which steps to take. Formal notational processing should be able to account for the observed effects of task structure variables. The present model suggests the importance of translation from imagistic configurations to formal configurations (by means of a process labeled FORMULATE in Fig. 6.1), and of using imagistic processing to motivate steps in formal notations. Thus, not only should the writing of equations for a verbal problem be influenced by the solver's visualization of the problem, but also the solving of the equations should often be done with reference to the meanings of steps taken on the imagistic level (by means of the process called INTERPRET in Fig. 6.1).

PLANNING LANGUAGE

Planning language includes heuristic processes used by problem solvers. It stands as a metalanguage with respect to the other languages, in that it not only controls the choice of steps within them but can also modify their rules of procedure; the process labeled CONTROL in Fig. 6.1 is envisioned as carrying out these two functions. The effects of heuristic behavior variables characterizing problem tasks are to be accounted for by planning language processing.

The main point I wish to emphasize here is that planning language also has a recursive capability, so that the heuristic processes that comprise it can act not only on domains from the other language systems, but also on the domain of heuristic processes. For example, a student may wish to solve a word problem in algebra. First she attempts the use of "trial and error," substituting values for the variables and trying to successively approximate the solution. Evaluating this method as unsuccessful after several trials (perhaps the solution involves fractions), she decides to write algebraic equations and solve them formally. On the domains of formal and imagistic processing, the student first uses trial and error, then equation solving; but on the domain of planning language, the student *also* uses trial and error, each trial being itself a heuristic process! It may be conjectured that the immense diversity observed in heuristic problem solving can be understood by means of relatively few processes, together with the capability of each process for acting on the others. This brings into play all the beautiful complexities of "level-crossing" and self-reference discussed, for example, by Hofstadter (1979). Resolved also is the problem of avoiding an infinite regress on the executive level, so that higher levels of control are not needed to explain planning steps.

EDUCATIONAL IMPLICATIONS

The model's structure has intentionally been kept simple enough that its main features can be communicated to teachers. The next step is to develop teaching objectives with explicit attention to all four of the language systems we have used to model problem-solving competency, and to the exchange of information among them. Whereas most school mathematics instruction is now devoted to the notational level, and some to the verbal level, little is devoted to heuristic planning or imagistic processing. Perhaps many of the failures of the "new math" curricula can be understood as caused by misunderstanding (by teachers or authors) of imagistic and heuristic objectives intended by mathematicians. For example, in the teaching of bases other than ten, a worthwhile mathematical objective could be improved visualization of the meaning of place value (imagistic level); another could be improved ability to discuss, compare, and modify conventionally adopted notation in mathematics (planning level). Few would have argued for a wide need to teach new mechanical algorithms for computation in various bases or for conversion between other bases and base ten (formal level). Perhaps predictably, algorithmic objectives tended to supersede the imagistic or heuristic objectives, because the latter were rarely made explicit. In consequence, much of the value of the new curriculum content has been lost.

We thus need to formulate problem-solving activities suitable for stating and achieving teaching objectives at all four language levels. Space permits but one example. In a well-known problem, "You are standing at the bank of a river with two pails. The first holds exactly three gallons, and the second exactly five. The pails are not marked for measurement in any other way. Find a way to carry exactly four gallons of water away from the river." Here the first stated objective should be construction of a nonverbal representation of the problem situation, and manipulation of it. Some children may experience anxiety when the problem is posed, because no previously taught algorithm applies. It may therefore be helpful to postpone posing the goal and instead describe the problem situation and ask: "What can you do? What do you see?" Children may be encouraged to imagine pouring water from pail to pail and asked about each step, "What happens if you do that?" Once a firm imagistic representation has been constructed, the goal of obtaining exactly four gallons can be introduced. Some students may generate many mathematical steps through visualization, but the next objective should be construction of a formal representation. Children may be guided to represent a situation by a pair of numbers and move back and forth between the imagistic level and the formal notation as steps are taken. Achievement of the goal state through standard "working forward" techniques can be a third objective, but the activities should not end there. On the planning level, objectives can be formulated and achieved that include the search for multiple solutions (this problem has two distinct solution paths), working backward, and problem generalization (discovering the possibility of obtaining a whole class of problems by changing the capacities of the pails). It is my hope that the model described here will help teachers employ such activities in meaningful, objective-oriented ways.

ACKNOWLEDGMENTS

I am indebted to A. Schoenfeld, E. Silver, and J. P. Smith for their interesting and provocative comments on the topic of this chapter.

REFERENCES

Bobrow, D. (1968). Natural Language Input for a Computer Problem-Solving System. In M. Minsky (Ed.), *Semantic Information Processing*. Cambridge, Mass.: MIT Press.

Goldin, G. A. (1982). Mathematical Language and Problem Solving. *Visible language 16*, 221–238.

Goldin, G. A. & C. E. McClintock, Eds. (1979). *Task Variables in Mathematical Problem Solving*. Columbus, Ohio: ERIC Clearinghouse for Mathematics, Science, and Environmental Education. Repr. (1984), Philadelphia: The Franklin Institute Press.

Hofstadter, D. (1979). *Gödel, Escher, Bach: An Eternal Golden Braid*. New York: Basic Books.

7 Representation and Understanding: The Notion of Function as an Example

Claude Janvier
Université du Québec à Montréal

We rarely mention the word "understanding" in our work because it encompasses so much that like an "iceberg" it conceals more than it shows. In order to relate understanding to representation, we now give a few features of understanding.

UNDERSTANDING

1. Understanding can be checked by the realization of definite mental acts. It implies a series of complex activities.
2. It presupposes automatic (or automatized) actions monitored by reflexion and planning mental processes. Therefore, understanding cannot be exclusively identified with reflected mental activities on concepts.
3. Understanding is an ongoing process. The construction of a ramified system of concepts in the brain is what brings in understanding. Mathematics concepts do not start building up from the moment they are introduced in class by the teacher. This well-known tenet is not easily nor often put into practice in day-to-day teaching.
4. Several researchers attempt to determine stages in understanding. We incline to believe that understanding is a cumulative process mainly based upon the capacity of dealing with an "ever-enriching" set of representations. The idea of stages involves a unidimensional ordering contrary to observations.

REPRESENTATIONS

We think it is worth making a distinction between representation on the one hand and symbolism or illustration on the other. Let us use Davis' et al. (1982) definition, which exemplifies what we mean:

> A representation may be a combination of something written on paper, something existing in the form of physical objects and a carefully constructed arrangement of idea in one's mind.

A representation can be considered as a combination of three components: symbols (written), real objects, and mental images. We believe however that verbal or language features are equally predominant because they are the links in between those elements. We assume also that one can find representation without a real object component. In order to show that such a subtle and intricate distinction brings about some payoff, we use it with the concept of function.

WHAT IS A FUNCTION?

Can a single definition encompass the rich meaning of such a notion. In order to simplify the algebraic or formal treatment of this concept in an axiomatic framework, contemporary mathematics has divised definitions of function based on the notion of cartesian product. For example, it can be viewed as a triple of sets (A, B, C) where $C \subset A \times B$ such that if (a_1, b_1) and (a_1, b_2) belongs to C than $b_1 = b_2$.

SEMANTIC DOMAIN

The idea of representation helps us in distinguishing several facets of the concept of function. With Freudenthal (1983), we believe that behind the general idea of function lies many basically different objects. Freudenthal uses a term equivalent to phenomenological status whereas we prefer the expression semantic domain. In fact, we both claim that even though we can define transformation, variable, sequence, permutation, and isomorphism within the framework of function, those notions remain substantially different in the sense that most forms of reasoning involving each of them are substantially different.

Let us scrutinize two semantic domains. When a function is envisaged as a *variable,* the role of the domain is often played down if not totally disregarded. The domain is implicitly used, necessarily ordered, and very often dense. The nature of the variation is stressed. Mental images related to it and typical verbal descriptions are closely connected to the primitive notion of variation and continuity. In fact, we see a variable changing and our concept of *variable is this capacity of the mind to characterize this change.* The continuous cartesian graph is then the natural illustration of a variable. As history tells us (see Youschkevitch (1979)), the use of a curve seems to be a prerequisite for the construction of this semantic domain, and it also seems to branch out from a rejection of the proportional change as the unique model of variation.

Functions as a transformation require more ''intellectual efforts'' to deal with because they cannot be conceived without some reference to the domain.[1] Mental images of the representation of this semantic domain must involve the domain[1] and co-domain[1] (source and image)[1] at the same time. Any illustration must suggest the typical essence of a transformation showing initial and resulting objects. The idea of invariant seems to be intrinsically linked with this semantic domain. Geometric transformation other than simple translations or rotations are good examples.

SEMANTIC DOMAIN AND REPRESENTATION

We introduced in Janvier (1980, Chapter 3) the translation skill table that shows translation in between two of the following modes of illustration: graphs, formula, table, verbal description. We insisted on the need for direct translations that are rarely taught in class.

We now believe that this conception must be widened. We think that it is better to use the word ''schematization'' (which sometimes may be an illustration) and use the word representation with its more general meaning (as in Davis, et al. 1980). A translation between schematization is then performed *within* a representation. By analogy, a representation would be a sort of star-like iceberg

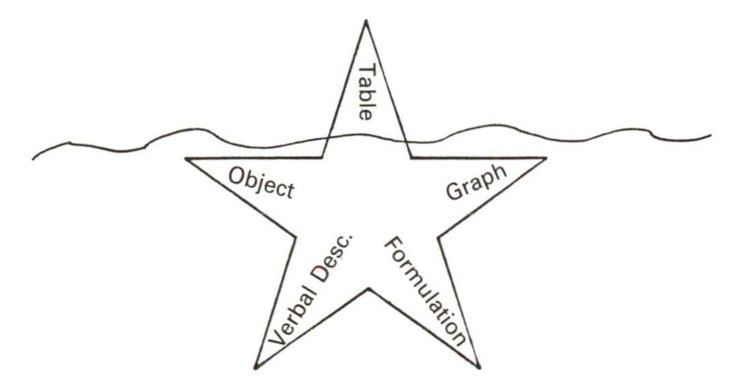

that would show one point at a time. A translation would consist in going from one point to another. This description of a representation has the advantage of insisting on the global and ''inseparable'' character of a set of schematizations.

---

[1]Mathematics meaning.

A MAJOR DIFFICULTY: THE CONTAMINATION COMING FROM CLOSE SCHEMATIZATIONS

A usual mistake when working within a semantic domain consists in transferring features of one schematization to another. We give three examples related to the idea of variables.

1. Contamination: verbal → formula. Clement and Kaput (1979) tell us about students' difficulty with the following problem: "At a certain university, there are six times as many students as there are professors." As they write: "25 to 30% of freshman engineering students write $6S = P$. This percentage go to over 50% when a nontrivial ratio is used." In addition to the wrong transfer of the linguistic form "Six times more students" to "$6S$," it is worth mentioning how "S" becomes student rather than the number of students as "g" is usually used for gram.

2. Contamination: graph → picture. Let us recall briefly the racing-car test item that we introduced in Janvier (1978). It showed that often pupils tend to see into a graph a total or partial picture of some situation involving the variables with which they deal.

3. Contamination: verbal → graph. We have very often noted that children have difficulty in "expurgating" the expression "grow fast" from the idea of being tall that it insidiously contains.

CONCLUSION

We conclude this chapter in suggesting the formidable distance between the richness of the concept of variable and the idea of function as too often presented in textbooks. Here is a function:

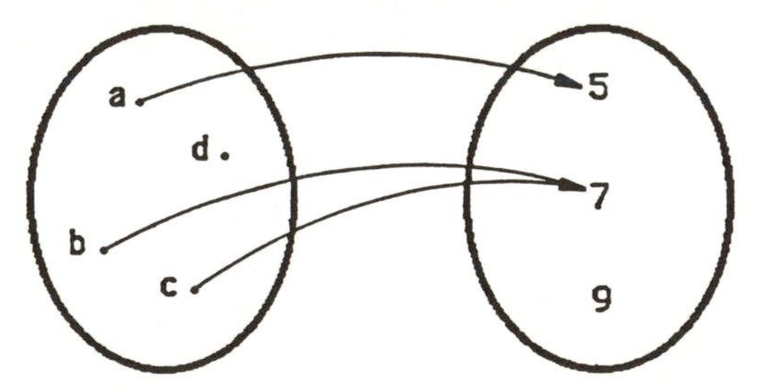

a) Give its domain, codomain, and range.
b) What is the image of b?
c) Draw its cartesian graph.

It seems that someone has forgotten something somewhere!

REFERENCES

Clement, J., & Kaput, J. J. (1979). Letter of the editor. *The Journal of Children Mathematical Behavior, 2* (2), 208.

Davis, R. B., Young, S., McLoughlin, P. (1982). *The roles of "understanding" in the learning of mathematic*. Urbana/Champaign: Curriculum Laboratory, University of Illinois.

Freudenthal, H. (1983). *Didactical phenomenology of mathematical structures*. Reidel Publishing Company. Dordrecht.

Janvier, C. (1978). *The interpretation of complex cartesian graphs representing situations: Studies and teaching experiments*. Doctoral Dissertation, University of Nottingham.

Janvier, C. (1980). Translation processes in mathematics education. *Proceedings of the VIth PME Conference*. Ed. R. Karplus. Univ. of California. Berkeley. 237–242.

Youschkevitch, A. P. (1979). The concept of function up to the middle of the 19th century. *Arch. Hist. Exact Sciences,* 16-1. 37–85.

8 What Do Symbols Represent?

John H. Mason
Open University
England

> I am a part of all that I have met;
> Yet all experience is an arch wherethro'
> Gleams that untravell'd world, whose margin fades
> Forever when I move.
>
> —from *Ulysses* by Tennyson

Distinctions between the symbol and the symbolized, the signifier and the signified have been drawn by many people in the fields of linguistics, semiotics, and mathematics. These notes are yet another attempt to bring to articulation something of the experience of using symbols in mathematics. Instead of dwelling on distinctions, an attempt is made in these notes to focus on symbolizing as a process. A metaphor is developed in which a symbol is seen as a door between two worlds, the material world of physical objects and the virtual world spoken of by Tennyson. Symbolizing is seen as a passage between worlds, either by trail blazing, or by following someone elses markings. My approach is tentative, and by means of mathematical examples. All assertions should be taken as conjectures, as attempts to speak the unspeakable.

PALPABILITY OF SYMBOLS

Example 1

> Consider the set of points in the plane that are twice as far from point A as from point B. After some mental imagery and trying out special cases, I resort to coordinates. Let $A = (a, 0)$, $B = (b, 0)$, $P = (x, y)$

Then $PA = K\,PB$ so $\sqrt{(x-a)^2 + y^2} = K\sqrt{(x-b)^2 + y^2}$ $(K > 0)$

Manipulating to get it in standard or recognizable form:

$$(x-a)^2 + y^2 = K^2(x-b)^2 + K^2y^2$$

From my experience I recognize a circle (as long as $K \neq 1$) because of equal coefficients of x^2 and y^2. Continue nevertheless:

$$x^2 - 2ax + a^2 + y^2 = K^2x^2 - 2K^2bx + K^2b^2 + K^2y^2.$$
$$x^2(1-K^2) + y^2(1-K^2) = K^2b^2 - a^2 + 2x(a - K^2b)$$

A and B are available to me to be chosen as two distinct points. By scaling and translation I can assume $a = K^2b$ without loss of generality, as long as $K \neq 1$. So

$$x^2(1-K^2) + y^2(1-K^2) = K^2b^2 - K^2b^4$$
$$= K^2b^2(1-K^2)$$

Thus $x^2 + y^2 = K^2b^2$, which is a circle centered at the origin with radius Kb. Choose b positive, so that $a = K^2b$.

For $K < 1$, the circle goes around A and not B.
For $K > 1$, the circle goes around B and not A.
As $K \to 1$, the circle tends to the right bisector of AB.

Where was your attention when reading my algebra? Fully on each symbol? Probably not. It is more likely that you slipped into an algebraic mode of processing, that you skipped over details noting merely the shape of the calculations. Now recall doing some algebraic calculations of your own. Your attention is partly on the details of accounting for coefficients, and partly on the shape of the whole, guiding yourself towards your goal. Often the goal is not algebraically specific, but rather a search for recognizable pattern or form in the symbols. A quadratic expression in x and y with equal coefficients for x^2 and y^2 is one such template associated with a circle. Most significantly, attention is not so much on the "symbols" as "through the symbols." Put another way, the symbols are not mere marks on paper but indicate or speak to entities that are almost palpable, almost substantial. When this happens, the symbols are no longer abstractly symbolic. Mason (1980) develops a spiral movement based on the enactive-iconic–symbolic distinctions of Bruner to account for the shift. The spiral describes an inner movement of increasing confidence:

from confidently manipulable objects/symbols,
through their use to gain a 'sense of' some idea involving a full range of imagery but at an inarticualate level,

through a symbolic record of that sense,
to a confidently manipulable use of the new symbols,

and so on in a continuing spiral. At any stage a question or surprise may induce movement back down the spiral to restore confidence and provide manipulable objects that can be used to investigate what is going on. Access to movement on such a spiral is one way of referring to the palpability of symbols, for the symbols are access to or windows on previous turns of the spiral. Movement up and down the spiral is what I mean by the gerund "symbolizing."

Whereas explicit attention is split between local and global detail, it is quite likely that there are accompanying images. These images can have aspects corresponding to any of the senses, though the most important for mathematics are probably the acoustic, pictorial, and kinesthetic. I have noticed that when I am doing complicated calculations I have either subvocal inner speech about moving terms around, or else a more distilled sort of acoustic response corresponding to various parts of the calculation. Kinesthetic images tend to be submuscular, involving actual or virtual muscular tension by analogy to subvocal speech. Pictorial and kinesthetic images are another way that I use to refer to symbols being palpable.

Whereas at some stages in example 1 some of the terms can be interpreted geometrically, particularly by following the Greek tradition and interpreting squared terms as areas, the palpability to which I refer does not usually consist of such detailed interpretation. Palpability has to do with potential interpretation, with confidence in the possibility of moving back down the spiral. Just as a food store manager can look at his inventory figures without having images of baked beans tins and cereal boxes lined up on shelves, so the mathematical symbols do not need to be specifically interpreted in detail to achieve palpability. We do not usually attend to interpretation in the middle of symbol manipulations unless we get stuck.

Now let us pause and reflect on what was happening in example 1. In one sense, the symbols are objects being manipulated. They are manipulated according to rules and with the intention of reaching a goal, particularly during the squaring and rearranging. The symbols have no salient geometric interpretation, yet they are nevertheless rich and substantial by virtue of algebraic experience; a "typical working out."

In another sense the symbols are artifacts resulting from recording perceptions on paper, mere vestiges of a complex inner experience. To someone experienced in algebraic manipulation, they speak to a rich inner world of sounds (did you not hear x^2?), images (points on a plane, . . .), and more vague "senses-of," perhaps with submuscular response. They act as blaze marks for a trail, indicating a route, and yet by their presence they extend the mental screen and make it possible to continue search for that route.

I suggest that these traditional "senses" are actually part of a single process that I denote by *symbolizing*. One reason for the ubiquitous distinction between signifier and signified may be that people confident with symbols no longer notice shifts in inner images that signal their participation in a symbolizing process. Instead they notice only their manipulative facility. Watching someone else move symbols around certainly gives the impression that the symbols are merely like counters in a game of Ludo or Monopoly—objects to be moved according to rules, but without significance. The purpose of example 1 was to show that the manipulator is doing more than shoving "counters" around in a formal game. The spiral gives one way of thinking about symbolizing as an act or process in which each person, trail blazer or follower, must participate.

Surface and Deep Structure

Skemp (1979) observes that attention is all too easily drawn to syntactic surface structure and away from semantic deep structure. This subverts the symbolizing process, perhaps obscuring it so thoroughly that teachers forget its role, and students never experience it explicitly. Some students only experience *other* people's algebra (like exercises in $2x + 3x$ and $x^2 + x^3$), without being encouraged to use algebra to express their own generality, to manifest their own inner perceptions in written form. This is because over the years the manipulative techniques have been distilled, and through examination pressure have become the focus of teacher and student attention. Practice of technique is important, but there must be some access to symbolizing so that if and when trouble develops students have recourse down the spiral to greater confidence and meaning.

The attraction of surface structure over deep structure is of course important in the movement up the symbolizing spiral, in which symbols become concrete, confidence inspiring, and palpable. It would be impossible to do computations if it were necessary to refer back constantly to meanings. Take for example the problem, Divide $2/3$ by $4/5$. People at all familiar with fractions simply invert and multiply, which is a surface structure response to syntax. Asked to provide a running commentary, most people find it very difficult to go deeper than syntactic manipulation rules. In such instances the technique must become divorced from meaning in order to become operational, and yet it seems to me important to be able to go back and regain confidence in why it works, by tracking back down the symbolizing spiral. Somewhere on that spiral there ought to be some images or "senses of" or stories that interpret and justify the technique.

This little example also illustrates for most people the struggle that is often encountered by students in moving from a vague "sense of," a seeing, to trying to say what they see, and then again to record what they can say. I suspect that most mathematicians feel they know what is going on when dividing fractions, but it is an interesting and difficult exercise to try to express a direct perception of why "it works." It usually takes several attempts, just to re-explore the problem and find an interpretation, and several more attempts to express it clearly. Even then it may not be so easy to write such an explanation coherently. Furthermore there remains a nagging doubt that there may be another way of thinking about it that is even more direct and "intuitive."

Another manifestation of the attraction of surface structure over deep structure is the propensity for teachers to concentrate on naming mathematical objects. It is often said that naming things gives power over them. However that power only comes from an inner movement or change in relationship to the thing named, a form of crystallizing, cohering, or precising. It's what you do with the names that counts. Just as a novice bird watcher can become so involved in the recognition and naming of birds that they fail to develop their observations into awareness of the habits and nature of particular birds, so being indoctrinated into standard definitions of circles, squares, and rectangles misses the point of mathematics. Classification is for the purpose of formulating theorems, not simply to achieve superficial classification.

The Voyage Metaphor

At the heart of most attempts to talk about teaching and learning lies a conveyance metaphor. This has been highlighted by Lakoff and Johnson (1980). Ideas are identified as objects that are conveyed from person to person: "Have you grasped my meaning?" "They haven't got the idea." "I tried to convey to them a sense of . . ."

Meaning is carried by words. Ideas are put *into* words, so that the meaning is literally carried. When words or symbols are received, it is assumed that the meaning is taken out of the container. This conveyance metaphor is ubiquitous, and remarkably subtle in its manifestations. Yet when looked at closely it seems almost absurd, because each person has to make his or her own meaning based on personal resonances with the words and symbols used.

I propose instead a voyage metaphor, which emerges from the considerations so far. Its essence is that mathematical thinking is like going on a voyage. Doing mathematics with others involves voyaging together, with all the agreements and disagreements of fellow travellers. Writing up mathematics is like reporting back, or writing a travelogue. Teaching mathematics involves being both a tour guide, and an old hand listening critically to fresh reports. The metaphor incorporates the trail blazing and window images for the role of symbols and emphasizes both the personal construction of meaning and the role of imagery in being mathematical. It also suggests that communicating mathematical ideas is like showing holiday snaps to a friend—you can see, hear, smell, etc. a rich background of context for each picture, whereas your friend has to make sense of what is shown. It can be helpful to talk directly from mental images of the context, describing the details in vivid terms. It is not helpful to gush a lot of generalities at our friends in an attempt to transfer our experience to them.

The voyage metaphor is not universally popular because it seems to imply some Platonic world of preexisting forms that mathematicians encounter, and this appears to conflict with the relativism arising from undecidable propositions and alternative geometries. I believe that this is a naive and simplistic version of Platonism, but this is not a suitable place to pursue such matters. Despite possible objections along these lines I invite you to consider whether you find any resonance between your experience and a "traveller's tales" perspective of symbolizing, and whether the metaphor informs aspects of symbols as representations and symbolizing as a process that are ignored in the conveyance metaphor. Some examples as fodder for such investigation follow shortly.

Some support for a voyage type of metaphor can be found in Skemp (1979), where he suggests that symbols can act as interface between "inner reality and outer actuality; inner realities of different people; conscious and unconscious levels of inner reality."

The third interface is another way of talking about trail blazing and resonance of meaning on the spiral. The second interface refers to the traveller's tales aspect of symbolizing, and the first connects with transitions from seeing to saying to recording, and movement along the spiral. Skemp's language corresponds to Tennyson's reference to two worlds, and hence to a voyage metaphor.

Here are some further mathematical examples on which to test the voyage metaphor and the spiral perspective.

Example 2

> Consider the following "standard" definition of continuity of a function F at a point ϵ:
> For all $\delta > 0$ there exists an $x > 0$ such that $|F(x + \epsilon) - F(\epsilon)| < \delta$.

The switching of roles between ϵ, δ, and x is deliberate, because it highlights the extent to which we depend on recognizing patterns in sequences of symbols. It seems closely analogous to the ability of pianists to sight read complex sonatas. They recognize whole chords at once, whereas the novice is looking at the individual notes. If the patterns are upset, as they were in the definition, then a taste of the novice's experience is available. As a topologist said: "I can't think about manifolds if they are not called X."

What did you notice about how you treated the symbols in example 2? Did you use pictorial images, either of the graph of a function in an interval, or of the usual definition laid out, and then match my version to yours? What can you say about the symbols—were you focused specifically on them, or were they more like shuttered or misty windows, which when opened or cleaned, again gave access into Tennyson's world? Continuity is quite naturally associated with pictures of functions, but what are students expected to make of group theory definitions such as $C_G(H) = \{g \text{ in } G : gh = hg \text{ for all } h \text{ in } H\}$

What internal images does that evoke? For me they are mostly acoustic and kinesthetic. How can students be awakened to such imagery of their own?

Example 3

> An equal number of red and blue points have been chosen in the plane, no three on a straight line. Is it always possible to match up the red points with blue points by straight line segments that do not cross each other?

Example 4

> Chickens in a coop rapidly develop a pecking order, in which for each pair of chickens one always pecks the other. If there is a chain of chickens of length at most K from chicken A and chicken B such that each chicken in the chain pecks the one below it, we say that A "K-dominates" B. For what K can we assert that there must always be a chicken that K-dominates all others?

Examples 3 and 4 illustrate some of the peculiar features of symbols and inner representations. A sensible place to begin tackling these sorts of questions is to try several particular examples, but this needs concreteness, some way to think about chickens and points that can be manipulated, and that displays what is important and ignores what is irrelevant—in short, a useful notation to represent what we see as being the essence of the problem.

While working with examples we are hoping/trying to see through the particular to a generality that will apply to all examples. This is what Hilbert meant by the use of generic examples (Mason & Pimm, 1984). Such seeing-of-the-general-in-the-particular is impossible if attention is captured by details of the examples. The way the particular is represented (whether internally or externally) must be as transparent or generic as possible. It must facilitate seeing generality. An analogy might be the childhood desire for X-ray vision, the fantasy being to be able to see through things. A good deal of mathematical thinking involves seeing the general in or through the particular and is very much like a childish version of X-ray vision.

Most people need to augment their inner vision or extend their mental screen when working on examples 3 and 4. Their pictures, notes, and jottings both aid concentration on detail and assist in keeping track of the overall route. Later they will act as trail blazes for reconstructing an elegent argument, which in turn functions as a blazed trail for readers. Readers will still be involved in symbolizing, but for them there are more or less helpful signposts along the way. Any resolution will certainly have to evoke images, or work with distilled representations of their essence, for instance with dots and arrows/edges.

Example 5

> Some spherical planets are in space. On some of the planets there are places from which no other planet of the system is visible. What is the total area over all the planets from which no other planet is visible?
>
> Crux Mathematicorum 1981:237.

Example 5 is rather different. It demands some internal sense or image of the planets, and some focusing on one planet to clarify the "regions of invisibility." A conjecture is not hard to come by, either by specializing to two dimensions and/or to just a few planets. Curiously, one line of attack for two dimensions, involving angles in a polygon, does not carry over to higher dimensions. It is once a conjecture has emerged and a convincing argument is being sought that the example points out some aspects of the role of symbols. It is remarkably hard to set down a convincing argument without resorting to all sorts of technical language, or verbose explanation that is hard for others to follow, and which does not actually represent the direct perception that led to the conjecture. To reach a satisfactory record of what is seen requires perseverence, a wish or need to sort it out. Not many students have this wish it seems. These examples remind me that mathematics is very often identified with symbols on paper, which is like identifying a traveller's tale with travel, or mistaking signposts for a journey. The voyage metaphor suggests:

taking opportunities for listening to students, to get them talking *because* transition from seeing to saying is a necessary part of the struggle towards meaning along the spiral, and because it is in the act of expressing to others that we often gain clarity for ourselves;

becoming aware of inner images and developing criteria for deciding whether to share these with students, and how

observing the effect of speaking directly from inner images as an accomplished traveller and raconteur, and of drawing attention to inner experiences as an important part of doing mathematics;

training students to notice their own inner images and to speak directly from them;

trying to enter a students world (mostly by listening) rather than trying always to drag them into our inner mathematical world.

Questions

The foregoing notes suggest a number of important questions:

1. How can you tell how substantial/palpable or how abstractly symbolic a symbol is for someone else? Asking them does not often give useful information!

2. Would talking aloud about feelings/images while doing algebra in public help or hinder the development of symbol sense in students?

3. How is symbol-sense developed? How can students best be weaned off physical and numerical specializations as their only refuge and domain of confidence?

4. Is the mere act of changing or augmenting a representation of more significance than the particular representation itself?

5. Does the voyage metaphor inform parts of teaching and learning that the conveyance metaphor cannot reach?

> "Tis to create,
> and in creating live a being more intense
> that we endow with form
> our fancy."
>
> —From *Child Harrold* by Byron

REFERENCES

Lakoff, G., & Johnson, M. (1980). *Metaphors we live by*. Chicago: University of Chicago Press.

Mason, J. (1980). When is a symbol symbolic? *For the Learning of Mathematics*, *1*(2), 8–12.

Mason, J., & Pimm, D. (1984, August). Generic examples: Seeing the general in the particular. *Jour. Educational Studies in Mathematics*. Vol 15(3) p 277–289

Reddy, M. J. (1979). The conduit metaphor, in Ortony A. (ed.), *Metaphor and Thought*, Cambridge Univ. Press, Cambridge.

Skemp, R. (1979). *Intelligence, learning and action*. London: Wiley.

Skemp, R. (1982). Surface structure and deep structure. *Visible Language*, *XVI*(3), 281–288.

9 Phenomenology and the Evolution of Intuition[1]

Andrea A. diSessa
Graduate School of Education
University of California, Berkeley

INTRODUCTION

In axiomatic mathematics it has long been recognized that beneath all the complex of definitions and theorems must exist a special layer that serves as foundation for the rest. Elements of this layer, undefined terms or axioms, etc., have the property of being primitive in the sense that they are not explicitly explained or justified within the system. Quite evidently the selection of this layer plays a fundamental role in determining the character of the system. In science as well, though somewhat less prominently displayed, selecting the primitives of a theory is an important and complex process.

This chapter is about primitive notions that similarly stand without significant explanatory substructure or justification. The system of which these primitives are a part, however, is cognitive, not a scientific theory or axiomatic system. We are after simple knowledge structures that are monolithic in the sense that they are evoked as a whole and their meanings, when evoked, are relatively independent of context. Because of this character we naturally find such primitives, like axioms, at the root of many explanations and justifications. In addition to explanation and justification, however, we are concerned with the important issue of control in human reasoning, not only what are the basic terms in which a situation is viewed, but how does one come to view it in that way? The context of exploring these ideas is physics, specifically the evolution and function of primitives in physics understanding, from naive to novice to expert.

[1]This is an abridged version of a chapter that originally appeared in *Mental Models* (Gentner and Stevens, 1983). More recent work by the author on this subject appears in *Constructivism in the Computer Age* (diSessa, in press).

Goethe gave an account of the development of scientific explanation that can serve to bring out the main features of the sketch of development in understanding that we propose. Goethe's claim is that scientific explanation begins with commonsense observation, a principal characteristic of which is its appearance as disparate and isolated special cases. What follows is a process of sifting through the cases, finding successively the more and more general and fundamental ones that serve as principles and explain the more special cases. In the end, one reaches the highest level that Goethe insists cannot be purely abstract but still must be phenomenological. In the following, Goethe (1978) describes these ultimate explanatory elements:[2] "We call these primordial phenomena, because nothing appreciable by the senses lies beyond them, on the contrary, they are perfectly fit to be considered as a fixed point to which we first ascend [in the process of finding what is fundamental], and from which we may in like manner, descend to the commonest case of everyday experience" (p. 72).

Something very much like primordial phenomena, we prefer the terminology "phenomenological primitives" (p-prims for short), are the central elements in the examples and analyses to follow. In brief the pattern we see is as follows. In the course of learning, physics-naive students begin with a rich but heterarchical (none being significantly more important than others) collection of recognizable phenomena in terms of which they see the world and sometimes explain it. These are p-prims. Some of these are compatible with the formal physics and are thus "encouraged," acquiring what we might call a higher priority than the others in terms of being readily used in physical analyses and explanations. In contrast to Goethe's philosophy of science, we do not propose that any of these are in themselves laws of physics. Instead they can serve a variety of cognitive functions in a physicist's knowing physics; for example, they can serve as heuristic cues to specific, more technical analyses. Typically, a process of abstraction including an expansion of domain of applicability accompanies this increase in priority and attachment to "textbook" concepts. In complementary manner, some p-prims lose status, very often being "cut apart," explained in terms of more fundamental, higher priority ideas.

What we wish to take from Goethe is the sense that direct experience in only mildly altered form can play a significant role in the understanding of "abstract" matters and his sense of continuity from naive to scientific apprehension of the world. The notion of evolution along the dimension of selecting some naively recognizable phenomena for more systematic and general application as knowledge structures will also be important. What we wish to leave behind are assumptions that phenomenology must be manifest in explicit science. Indeed, it appears in most of our examples that the work being done by p-prims is covert, perhaps necessarily so.

[2]An article along lines similar to this chapter, though not cognitively motivated, is Zajonc (1976).

At another level entirely, that of cognitive modeling, the choice of primitives in this chapter warrants comment. In particular, the process of recognition of phenomena is central. In addition to being motivated by the examples to follow, there is a significant body of experimental data that suggests recognition is a fundamental operation of the human cognitive apparatus. Recent work in facial recognition even suggests specific neurophysiological support (Carey & Diamond, 1980). Thus we consider it a good candidate for a "black box" to be used in theorizing about more complex cognitive activities. The points we make relate to what is recognized when, and how that contributes to physical reasoning; the process of recognition itself remains opaque.

The empirical basis for this work comes mostly from a series of in-depth interviews (about a dozen 1-hour sessions per individual) with four M.I.T. undergraduates taking freshman physics. The interpretations we offer are to motivate and explain the ideas we propose; they are not offered in proof. For these purposes, citations from protocols are paraphrased to increase clarity and brevity, and we also use informal observations made outside the study.

SPRINGINESS

M. was proposed the following problem: If a ball is dropped, it picks up speed and hence kinetic energy. When the ball hits the floor, however, it stops (before bouncing upward again). At that instant, there is no kinetic energy because there is no motion. Where did the energy go?

To a physicist, the analysis is straightforward: The ball and floor are mutually compressing on impact. That compression, just like the compression of a spring, stores energy in mechanical distortion. That is where the energy goes when the ball comes to a stop. The distortion, in fact, is what pushes the ball back up into the air on rebound.

M. was a quite respectable physics student whose analysis of the falling ball, gravitational potential energy gradually converted into kinetic, was clear. She had herself made the observation that a bouncing ball must stop at some point before its rebound upward. Further, she had an unequivocal commitment to conservation of energy and accepted the question about what happens at the bottom of the fall as a natural, even obvious one to ask.

M. quickly discarded gravitational potential energy as a repository for the "lost" energy and proceeded with the working hypothesis that the floor or earth must get the energy. As she appeared to be ignoring the role of springiness, the interviewer tried to prompt her by focusing on the rebound—the compression "pushing back"—thinking that a salient feature of springiness. "What causes the ball to acquire a velocity after it has stopped?" This involved her in elaborate rationalizations for the energy not staying in the floor. The gist of her clearest argument was that the floor is essentially an infinite mass (the earth), which in its resistance to moving could not permanently "accept" the energy and therefore "returned" it to the ball. She had enough sense of physical mechanism to be disturbed by her own almost animistic explanation but could not offer any other.

Seeing no progress in this line, the interviewer offered a straightforward explanation. "Think of a situation with a mass on the top end of a spring, where the mass and the spring fall to the floor." M. spontaneously continued the explanation; spring compresses, stores energy, pushes mass back up. Then the interviewer continued. "The ball is really the same thing, with the 'squishing' of the ball replacing the spring's action."

Although the mechanism of spring and mass was clear to M, she countered that, though a tennis ball might bounce in that way, she did not really believe bouncing could generally work like that. "What about a ping-pong ball? It doesn't squish."

The interviewer responded that it really did, but only a very small amount. However, if one took a strobe photograph and analyzed it carefully, one would see compression.

M. was undaunted. "How about a steel ball bearing?" The interviewer had to admit that depending on the materials involved, it might be the floor that did the squishing to absorb energy. But the principle would be the same.

M.'s final example to refute springiness was a glass ball on a glass plate that she knew to bounce very well—and yet glass would surely shatter rather than squish. Running out of time (and arguments) the interviewer terminated the session assuring that springiness was really there.

The next week M. was again asked about the bouncing ball. During the week the physics class had started kinetic theory, and she said she had a clearer idea where the energy was because of that; it must be in the internal kinetic energy of the molecules of the ball and/or floor! She did not mention springiness. (Though it is possible that energy could be transformed to internal kinetic energy of this sort, heat, that cannot account for bouncing. There is no mechanism to return the energy to gross motion.)

Rigidity and springiness are the p-prims at issue here. M. Does not see springiness in the ball bounce. Moreover, her counterexamples show this is no accidental oversight. She is used to thinking about "hard" objects as being rigid and appeals to her commonsense characterizations of everyday phenomena as self-evident justification: Steel is hard (rigid); glass is also hard and, in fact, responds to attempts to deform it by breaking. Within her intuitive frame there is little point in asking why steel is rigid and even less in asking for a justification of the applicability of the concept to the world at all. Rigidity is simply a property that she knows, by example, some objects have.

In contrast, physics experts have a much lower priority attached to the phenomenon of rigidity, although it will be assumed in restricted contexts, typically in making geometric calculations. This low priority is principled: Relative rigidity is an effect explained by much higher priority ideas such as forces (e.g., intermolecular electrostatic forces). Physicists have a strong particular sense for physical mechanism that includes the fact that looking at the world in terms of forces is fundamentally the right kind of explanation for physical phenomena.[3] A physicist views rigidity as irrelevant to any deep explanation of how things happen.

Springiness is another matter. It is not only consistent with the highest priority (Newtonian) physical ideas, it provides a convenient organizing conception that frees one from the necessity of always treating spring-like phenomena in terms of idiosyncratic situational details such as how and where exactly physical deformation is taking place. It serves as a macromodel that summarizes the causality (deformation → restoring force → rebound) and energy flow (deforming force drains energy into potential energy that is liberated as the deformation relaxes).

For the expert, springiness is a more powerful explanatory concept than rigidity, but it has definite limits. In the physical world there is no system that has exactly the properties needed to match the idealization, e.g., energy content dependent only on displacement (not velocity). Thus a physicist is not surprised if a numerical prediction made with the concept is in error by a small percentage. Even more, his understanding of how springiness relates to more fundamental concepts alerts him to circumstances when it will fail drastically. For example, a long steel rod bouncing on concrete involves wave motion internal to the rod rather than simply uniform compression. This means internal energy during the bounce does involve significant kinetic energy of the molecules, albeit in a coherent form rather than in M.'s "kinetic gas" image. An observation like this is sufficient to cause an expert to abandon macrospringiness and begin to construct a more elaborate special model.

Control of reasoning is an important issue here, and it is worth abstracting the function of springiness in expert thought with regard to this. By seeing the phenomenon of springiness in a situation, an expert brings to bear a default macromodel, sparing a great deal of situation-specific reasoning. The more "clever" the mechanism that cues springiness to recognition (triggering it only when it is a good approximation), the better off is the expert. But as we assume that the cueing mechanism is relatively simple so as to be quick, its pattern matching will not be perfectly reliably correlated with circumstances under which springiness will prove useful. Hence, although it has a relatively high priority, under certain circumstances it will defer, likely to higher priority concepts (in this case probably micromodeling on the basis of discrete masses and intermolecular forces or some continuum approximation thereof).

Implicit in these observations is a twofold structuring of the concept of priority. First of all, we must differentiate flow of control to and away from the use

[3]To explain how this sense is encoded is part of the point of this chapter. Our proposal, in brief, is that explanatory knowledge of a certain type has a particular place in a priority hierarchy.

of springiness. The former, which we call *cueing priority,* has to do with how likely the idea is to be profitable. Springiness is much more consistent with a Newtonian world view than rigidity, so in expert thought it will be used with less provocation than rigidity—it has a higher cueing priority than rigidity. Once cued, the resistance to abandonment is a second kind of priority that we call *reliability priority.* This second kind of priority is closer to being a proper technical sense for the notion of "more fundamental." Springiness, as we pointed out, is abandoned by experts relatively easily compared to such perspectives as provided by, for example, force and energy.

The second structuring of priority comes from the simple observation that specifics of the situation are relevant to deciding the priority of the use of an idea. In other words, priorities are context dependent. In this sense, it is even clearer that there is a difference between M.'s priorities as regards springiness and an expert's. M. certainly views a spring as springy, yet it does not occur to her at all to see a bouncing ball in terms of springiness. M.'s perception of springiness must change to become like an expert's specifically with respect to enlarging the set of contexts that cue the idea. In a similar way, reliability priority must change. Although it is not demonstrated by the protocol, M. does not know very much about the circumstances (context) under which even a spring should not be regarded as springy.

The context dependency of priorities allows us to make more apt descriptions of the class of "fundamental" ideas. The high reliability priority of conservation of energy is better described as "in a context where energy is applicable, one must heed conclusions drawn within its perspective." In contrast, conclusions drawn on the basis of springiness might be ignored even if it seems the concept is applicable, particularly if those conclusions are in conflict with those of a higher reliability perspective.[4]

It seems extremely plausible that the knowledge that structures reliability priority should regularly point to specific alternatives or refinements based on context. This kind of chaining of control pathways is important in later examples. For now we can illustrate the idea by hypothesizing that one way the context that cues springiness could be augmented is that another p-prim present in naive students, "bouncing," comes to defer very readily to springiness. Bouncing never comes to have a very high reliability priority but serves its function as a rapid heuristic control link to springiness. Because of its low reliability priority, an expert would be unlikely to appeal to bouncing in any

[4]The language used here, "drawing conclusions," "ignoring," and "heeding," is unnecessarily anthropomorphic and suggests a relatively elaborate reasoning process. In reality, of course, what goes on is likely unconscious and too quick and simple to be dignified by such high-level terms. We are after low-level control of reasoning mechanisms, and short of developing a technical vocabulary here, we rely on metaphoric use of these familiar but higher level terms.

explanatory way, and its effect might well be consigned to the "junk category" of intuition or "just knowing," e.g., that springiness is involved.[5,6]

The kind of development proposed here poses interesting pedagogical problems. Reorganization and change of function of knowledge structures seems less obviously amenable to external manipulation than "giving students new knowledge." The case in point is M.'s resistance to seeing a bouncing ball as a spring. How could one convince her that one should do so? In the absence of a more elaborate theory, we propose only that the process must be an extended one in which the coherence and success of the evolving new control system gradually compels reorganization of priorities.

M. gave additional force to her arguments against springiness by appealing to other well-known phenomena characterized in commonsense terms. Harder objects typically bounce better. If the "give" of objects accounts for bouncing, how is it that things with a lot of give, soft things like clay, don't bounce at all, and things with little give, hard things like ball bearings, bounce so well?

From a physicist's point of view, M. is making a mistake here that is common in naive and novice students—confusing softness (a small modulus of elasticity) with lack of reversibility (hysteresis) that causes smaller rebound. There is no necessary correlation between the two. In everyday objects with everyday forces, however, softer objects undergo greater deformation that, in turn, is often associated with more loss of energy, greater hysteresis; objects with more give generally bounce less well.

Though it is tempting to reify this conclusion (softer objects bounce less well) into part of a naive concept of springiness, M.'s protocol here showed a pattern of examples first (clay, steel, etc.) with conclusions following. This is in contrast to the issue of whether bouncing involves springiness where examples followed pronouncements and were apparently drawn up in an attempt to justify her impressions. Her mistaken understanding of the character of springiness would seem to be emerging from on-the-spot reasoning based on well-known phenomena. The following interpretation suggests itself: M. percieves a family resemblance among the various cases of objects giving when stressed, a naive p-prim that we might call "squishiness." In order to elaborate her own understanding of the phenomenon, she examines a few typical cases of it and concludes rebound decreases with softness. It is important that hysteresis, lack of rever-

---

[5]Cueing and reliability priorities are very general control constructs, and we do not mean to reserve them for phenomenology. But they seem particularly well adapted to explain control of reasoning where a large number of competing perspectives might apply. Motivated by some of the same considerations, Clemenson (1981) has developed a control system that one can understand as a particular version of structured priorities in the context of a computer-implemented problem solver.

[6]Incidentally, one sees an important coherence in the naive acceptance of a world including rigid objects and the acceptance of bouncing as a primitive that needs no mechanistic explanation.

sibility, is not salient for her, and that she does not pursue the question of *why* such a correlation holds. It seems plausible she might now simply remember the conclusion. But even if she does not, her vocabulary of phenomena and typical examples thereof means she would likely rederive the result on future occasions.

A final excerpt from a protocol points to intermediate stages in the change of function and reorganization of priorities on the road from naive springiness and rigidity to those of the expert. T. was given the following problem after demonstrating that he was clearly beyond M.'s belief in rigidity and unrefined sense of springiness: A pencil can be balanced horizontally on a finger. The interesting fact is that the configuration is stable under small perturbations; when pushed, it returns to horizontal. Why?

The problem is tricky because it is very tempting to assimilate it to a "standard balance" geometry, an object pivoting on its center of mass. The pencil pivot is not only not at the center of mass but even moves as the pencil rolls back and forth on the finger. Applying standard methods of energy, etc. with the false assumption of pivot at center of mass does not predict stable balancing.

Some students take stable balancing as a primitive phenomenon, "that's the way balancing works," and never bother using textbook physics on the problem. At best they appeal vaguely to sanctioned principles like symmetry. T. was also more advanced than that. Though he fell into the trap of the false geometric model, that slip allowed a good glimpse of his rather well-developed, though not quite expert-like, notion of springiness.

T. started with a fairly long, of course unsuccessful, attempt at using forces, torques, and detailed energy considerations. Searching for another way to look at the problem he focused on the process of pushing one end down a bit, thus doing work on it, and then on the conversion of the "stored" energy into kinetic energy as the pencil is released to swing back to its horizontal position. T. described the situation as "like a spring." More than a description, he proposed that analogy as an explanation, we would say appealing to springiness as a primitive explanatory phenomenon. When queried about how comfortable he was with that as a physics explanation, he indicated that he really was a bit reluctant to make it and suspected there might be a better way to explain it.

T. has evidently abstracted springiness to the level of macromodel for energy storage and restoring. We also would interpret the fact that he thinks to use it in this nonspring situation as indicating he has expanded, perhaps too much, the contexts in which he appeals to the idea. Possibly T. directly recognized springiness in the pattern of response of the system to perturbation, or possibly at the abstract level of energy storage. Either way, this suggests his cueing priorities have advanced, and he is not just bringing springiness to bear without first seeing its relevance. His tentative use of the concept as an explanation also indicates a substantial reliability priority, probably more than a naive person, but that priority is clearly not structured like an expert's. An expert would be looking for a deformation to decide whether springiness was relevant.

In summary, we see the development of the p-prim springiness from naive to novice to expert in the following way. The everyday phenomenon of springiness gradually becomes abstracted and modified from a root "squishiness" to a model[7] of an idealized spring-like causal and energy storage mechanism that helps physicists avoid unnecessary, detailed, situation-specific considerations. Cueing mechanisms to springiness become changed, specifically broadening to include all bouncing, and specifically made narrower to exclude large hysteresis examples like clay. An expert will use the phenomenon in some cases as a primitive explanation but also must defer to other ways of viewing the situation if he is being careful, i.e., if he is requiring his analysis involve only notions with the highest reliability priority.

OHM'S P-PRIM

Think of a vacuum cleaner whose intake nozzle you hold in your hand. If you put your hand over the nozzle, will the pitch of the sound you hear from the motor go up or down?

There are three common classes of answers to this problem (in this case we do not deal with expert solutions): (1) Some subjects do not understand enough of the mechanism of sound production to answer reasonably. Typically an analogy is made to the tuning of a musical instrument and pitch is assumed to be related to length of pipes, etc., instead of directly to the speed of the motor. This class is of no interest. The other two classes draw opposite conclusions about the change of pitch, interestingly for essentially the same reason; (2) The second class has the pitch (speed) of the motor reduced. This is attributed to an interference with the action of the motor by the hand and is occasionally explicated with reference to examples like an electric drill slowing down when the bit is inserted in wood. Almost never is there any explanation beyond that; (3) Others answer that the motor must speed up—the motor is being interfered with, and the motor must "work harder" (or some equivalent phrasing) because of the interference.

We explain both of these answers using a primitive explanatory phenomenon we call Ohm's P-prim. (The reason for the terminology becomes apparent later.) It comprises three elements, an impetus, a resistance, and a result. The impetus acts through the resistance to produce the result somewhat as schematized in Fig. 9.1. These elements are related through a collection of qualitative correlations such as an increase in impetus implies an increase in result, an increase in resistance means a decrease in result, etc. Recognition of the phenomenon in the

[7]In contrast to other kinds of models, this type is not constructed on the spot; its usefulness is in serving as an archetype. Furthermore, it is essentially monolithic and not generally open for inspection in terms of more primitive elements. Some researchers call similar knowledge structures "schemata."

FIG. 9.1. The Ohm's P-prim.

vacuum cleaner situation is compellingly simple. The motor provides a model impetus, the hand a clear blocking interference, and the spinning of the motor coupled to the flow of air is the result. Class two answers appeal directly to Ohm's P-prim as an explanation for the slowing of the motor. The features of the situation are so obviously those of Ohm's P-prim that some subjects report remembering that the pitch goes down. (In fact, it goes up.) Like the low-level processing that causes visual illusions by responding to cues that are not always associated with genuine features of an image, the p-prim can cause a "conceptual illusion," so that the person perceives interference where there is none.

Class three explanations draw the correct conclusion, that the motor speed and pitch go up, based on the same Ohm's P-prim illusion but adding an anthropomorphic feedback loop: as the interference (resistance) increases, the motor must increase its effort as if to make up for increased resistance. Anthropomorphism of this sort is frequently offered by physics-naive people as a primitive explanation, but its priority in the context of purely mechanical situations drops quickly as technical sophistication increases. In many cases, it appears that this anthropomorphism is forced as a rationalization for what they remember, speeding up, in view of the evident Ohm's P-prim that predicts slowing down. Some of these subjects admit sensitivity to anthropomorphism's defects. Some even spontaneously wonder out loud how the motor "knows" what's going on, or comment on the lack of a mechanism for a motor to "want" to produce some result. To support our contention that class three explanations are composite Ohm's P-prim plus anthropomorphism, some subjects after questioning and rejecting their own anthropomorphism are left with Ohm's P-prim and begin to question their remembrance that the pitch goes up.

We take Ohm's P-prim to be a very commonly used, high-priority p-prim. Contexts of application range from pushing harder in order to make objects move faster, to modeling interpersonal relations such as a parent's offering more and more encouragement to counter a child's offering increasing resistance. Although we think it provides a qualitative model and, at early stages in learning, a self-contained explanation for many more formal concepts in physics, we focus on only one, Ohm's Law, $I = E/R$: the current flow in a circuit, I, is proportional to the voltage (also called potential), E, and inversely proportional to the resistance, R. The interpretation of $I = E/R$ in terms of the Ohm's P-prim is straightforward and a good match with other characteristics of the physical quantities. Voltage is an independent impetus that causes a dependent amount of current to flow as a result. Resistance modulates current flow in the appropriate way.

This interpretation of $I = E/R$ is so natural that it is easy to mistake it for understanding the equation. But, though physics texts implicitly and sometimes explicitly invoke it for its mnemonic effect and because proper attachment of the Ohm's P-prim to the equation makes qualitative reasoning about varying quantities quick and easy, still the interpretation is only an interpretation. What follows is a typical textbook invocation of the p-prim and a warning that it (particularly its causality) is not inherent in the physics (Ford, 1968): "A simple way to look at the law is to picture the potential as the "motive power" (the British word for potential, "tension" is suggestive), the current as the resulting effect. Doubling the potential across a circuit element causes a doubling of the amount of charge that flows through it in one second. This assignment of cause and effect, of course, is only an aid to visualization, and has no deep meaning" (p. 551).

The author deliberately uses the metaphor "motive power" to enforce the impetus interpretation of potential, and he describes current as the resulting effect before warning that this too is metaphorical. His remarks also remind us that these associations were almost certainly in the minds of the scientists who first discovered such laws and then codified their interpretations in technical terms chosen to convey the sense, like *tension* and *resistance*.

In terms of our theory sketch we can summarize as follows: The Ohm's P-prim becomes profitably involved with the physical Ohm's Law as a model of causality and qualitative relations compatible with it. It initially serves novices as an explanation of the law and comes to be cued in circumstances where $I = E/R$ is applicable providing rapid qualitative analysis. With experts the p-prim is still useful[8] but has low reliability priority; it will not be identified with the law or used to explain it (except as a deliberate analogy).

A NOTE ON ABSTRACTION

We have said little about the process of "abstraction" through which naive phenomena become changed to serve as expert p-prims. So as not to leave this box entirely black, consider a thought experiment. A novice has been learning

[8]One might think experts would drop such interpretations, except for pedagogical purposes. On the contrary, at least in some circumstances they invent more of them! Electrical engineers often speak of a resistor as a kind of transformer that converts current flow into voltage. "A known current causes a potential drop IR in flowing through a resistance." This causality, which is the reverse of the Ohm's P-prim interpretation, is equally metaphorical. It is assumed when convenient but ascribed no deep meaning.

about potential energy, perhaps even in the context of springs, and decides to run through a squish–unsquish cycle in his head. He imagines his hand slowly pushing the coils of the spring together realizing his agency and the work he is doing on the spring—pumping energy into it. Simultaneously he feels that the spring "wants" to cause his hands to fly apart. The energy is not gone and in releasing the spring one can feel the work it is doing, now on his hand.[9] (M. probably had gotten this far; she understood springs. But as far as priorities, her "squishiness" dominates in contexts other than literal springs.)

Notice how the elements of the interpretation, in particular the set of features to be attended to, are mostly drawn from common sense, and yet the combined effect is to serve as the model of causality and energy storage that is the function proposed for the expert p-prim. The thought sequence binds together in an appropriate way the elements of previous knowledge that serve as basis for the interpretation. The structure of that combination is the new element for the student. That structure is locally justified because it is seen merely as a description of a known phenomenon, the action of a spring.

The abstraction going on here is essentially the selection of a conventional interpretation of an everyday event. It is abstract only in that the "actors" in the phenomenon (e.g., person, spring) are subordinated to become instances fulfilling their "role" in the interpretation (e.g., cause of squishing, thing that deforms).

SUMMARY AND CONCLUSION

P-prims are relatively minimal abstractions of simple common phenomena. Physics-naive students have a large collection of these in terms of which they see the world and to which they appeal as self-contained explanations for what they see. In the process of learning physics, some of these p-prims cease being primitive (and are seen as being explained by other notions), and some may even cease being recognized at all. But many become involved in expert thought in very particular ways. We have tried to highlight two ways in our examples: p-prims serve as elements of analysis, we might say models, which partially explain and provide rapid qualitative analysis for similar but more formal ideas; the recognition of p-prims can serve as a heuristic cue to other, typically more formal, analyses.

In becoming useful to experts, naive p-prims may need to be modified and abstracted to some extent. In particular they will need to be recognizably applicable to a different, usually more general, set of contexts. Just as important, they

[9]Though this thought experiment was fabricated, in the time since this chapter was originally written several students have recounted strikingly similar episodes. These should appear in future publications.

will not often serve the same naive role, self-explanatory analysis, and must defer to other analyses for that purpose. We have developed the idea of structured priorities to refer to the complex of mechanisms that determine when a p-prim is recognized in a situation and when it is abandoned from consideration, perhaps in deference to some particular other view of the situation. This theory-sketch highlights a difference between novices and experts, indeed between common sense and scientific reasoning, which is not so much the character or even content of knowledge, but rather its organization. Experts have a vastly deeper and more complex priority system. Physics-naive people's knowledge system is structurally incapable of supporting any strong, principled commitment to a particular interpretation of a physical phenomenon.

On a higher level this theorizing is an attempt to understand the potentially pervasive and fundamental function of recognition in human cognition of a large collection of events as like some particular one. In the same vein, the notion of p-prim suggests how another relatively primitive operation, the remembrance of an interpretation of some event such as the compression and relaxation of a spring, complete with selective focus on particular features, can have important and far-reaching effects when integrated into an elaborate control system that "knows" when and when not to see a situation as like that particular event. Finally, we have suggested that it is important to know about a naive person's repertoire of p-prims and how easily their priorities can be molded when considering how one should explain advanced notions in such a way as to be understandable at the student's level, but just as important, in such a way as to develop naturally into expert understanding.

ACKNOWLEDGMENTS

I would like to thank Melinda diSessa, Jim Levin, and Dedre Gentner for careful reading and specific comments on previous drafts that greatly affected the final form of this chapter. Others whose comments have helped me clarify these ideas include Jeanne Bamberger, Bob Lawler, and John Richards.

REFERENCES

Carey, S., & Diamond, R. (1980). Maturational determination of the developmental course of face encoding. In D. Caplan (Ed.), *Biological studies of mental processes*. Cambridge, MA: MIT Press.

Clemenson, G. (1981, January). *A case of dependency directed problem solving*. (Doctoral thesis). Cambridge, MA: M.I.T. Department of Mathematics.

diSessa, A. (in press). Knowledge in pieces in G. Forman (Ed.), *Constructivism in the computer age*. Hillsdale, N.J.: Lawrence Erlbaum Associates.

Ford, K. W. (1968). *Basic physics*. Waltham, MA: Blaisdell Publishing Co.

Gentner, D., and Stevens, A. L., eds. (1983). *Mental Models*. Hillsdale, N.J.: Lawrence Erlbaum Associates.

Goethe, W. v. (1978). *Theory of Colours* (Trans. C. L. Eastlake). Cambridge, MA: M.I.T. Press.

Zajonc, A. G. (1976). Goethe's theory of color and scientific intuition. *American Journal of Physics, 44*, 4.

II REACTIONS TO THE PAPERS

10 Excerpts from the Conference

A MODEL OF RESEARCH
Richard Lesh

You can also set out to discover children's cognitive characteristics. An example of work to do in this area: individual differences, where you would make some generalizations about a child. He is impulsive, say, or reflective. To my way of thinking, that is one step too far in the way of explaining things across many tasks. I see children who are sometimes impulsive in a problem-solving setting, sometimes reflective. The explanation that I want to come to is when is it beneficial or when has it to do with the underlying concepts. So, you can study either how children develop or how ideas develop in children and it makes the world of a difference about the kinds of generalization that you form. I am uncomfortable with assigning labels to a kid. I don't mind assigning labels to a kid's use of a particular idea. That's the kind of analysis that I like and I think that it characterizes what has occurred in mathematical education research in the last 10 years. There are psychologists doing either task analyses or analyses of children. In the middle, there is a cluster of people studying the way ideas evolve in children. That I think is what mathematics educators uniquely have to contribute to research in that domain.

ARE THEY PRIMITIVE?
John H. Mason

My question to Andy (diSessa) was: Can you give me examples of mathematics p-prims? Before I do I would also like to raise the question of primitivity. I am reminded here of my interpretation of Bruner in which I have the sense of a spiral movement; it isn't linear. So one would do his work with some material, some apparatus if you like. It leads to some sort of inner ideas, inner awareness, that eventually become articulated and come out perhaps sort of looking like prims, like phenomenological observations. Those articulations become refined and become the objects to be manipulated later. I have the feeling that this spiral comes down quite far as well. It doesn't just start and build up. So I've been thinking about things like multiplication by minus one. A question you might ask yourself is what came up inside you when I said multiplication by minus one. When I said it you became aware that you have access to certain things. But I'm not sure that they are primitive.

CLUSTER OF REPRESENTATIONS
Richard Lesh

Kids go to the world and build a cluster of several representations that interact among themselves and with the problem; and they go from that cluster of representations and don't map back to the real situation. They often take that (cluster) too as a given map and to another. So the solution process doesn't look like this: real world→model→map back. It looks like: real world→several things map to maybe another loop over here and maybe coming back. These things happen frequently, if you watch children solving this sort of ''realistic'' problem. Kids are having problems with trying to get things into a single coherent representational system so they can process them.

FORMAL NOTATION
Andrea A. diSessa

Let me make a couple of comments about formal notation. I wonder if physics is essentially different from mathematics in that Gerry (Goldin) says that 95% of school mathematics has to do with formal notational systems. I certainly don't think that physics is anything much like that. Maybe there's an essential difference between physics and mathematics, but maybe the difference is in our emphasis in selecting what is fundamental to understanding, or maybe the difference is in the level of description we are chosing. In the latter category, I would like to make an important point. I would say that 95% of the formal notational processing is imagistic. How do people encode their understanding of what to do with these symbols? It's quite clear to me that that has significantly to do with spatial metaphores and things like ''moving objects around.'' All of this is of the same kink, at the same intuitive imagistic level on which a lot of semantics runs. It happens to be directed at a different class of things, symbols, but that does not mean that it is essentially a different kind of knowledge, that it is organized in any essentially different way.

IMAGISTIC CAPABILITY
Claude Janvier

When we discuss solving a problem, I do remember talking to someone about the following problem: Three objects cost two, how much do six objects cost? That person said that it is easy because it is cross-multiplication. Well, we could say that if it is twice more so it will cost twice more. This reaction shows that in school you have a selection of strategy being done. If cross-multiplication is done all the time, school teaching or textbooks sort of reduces the imagistic capability of children because the strategy of doubling that is basically different and more visual is systematically played down.

INTUITIVE EPISTEMOLOGY IN REFERENCE TO GOLDIN'S MODEL
Andrea A. diSessa

Let me talk about some other things that I think ought to go in that functional category (planning). I think that something called intuitive epistemology is a good candidate for something that's important and very nearly in this planning language level that is still left out of this particular description. The feelings of understanding or of not understanding I think are vitally important to planning and controlling your activities, yet they don't have built into them a suggestion of what you should do. So there is a certain class of knowledge that is vitally important for deciding what you should do (if you should do anything) but doesn't have this characteristic of control.

MESSY CONTEXT
James J. Kaput

I am suggesting that we need to change the focus and look at what the characteristics are of particular representations that make them effective in some cases and ineffective in others. That is going to be hard as Lesh has been pointing out, but people, particularly students, don't use representational systems in a pure way. Their understanding of the syntax of a particular representational system is in itself distorted. So, it is one thing to come up with a general theory describing representational systems and describing the characteristics of the representational systems. It is another thing to apply that theory in a messy context like applied problem solving or even to concept development.

MORE THAN COMMONSENSE OBSERVATION
Ernst von Glasersfeld

Another term that is used in Andy's (diSessa) chapter is *common sense observation,* which I think is the source of the original p-prims. It consists of sensory–motor records but it is more than that: P-prims are already attempts at explanation. They are not just records of an experience. He then says that they have been extracted from a variety of situations and these situations again are the subject's own operations and sensory experiences. The one thing that puzzles me is that there are certain phenomenological notions that I would call p-prims that have to be completely disregarded and thrown away if you begin to do physics. So I have a little difficulty about the word phenomenological in that context, because the word could refer to everything one experiences in everyday life as well as to the physicist's experiments—both are phenomena, i.e., items of someone's experiential world—and I'm not sure how I am to separate them.

MULTIPLE EMBODIMENT PRINCIPLE
Claude Janvier

I was very much surprised the first time, 10 or 15 years ago, I read Dienes' principles; more particularly the one stating that an ideal method of learning mathematics would be to use several representations of the same object.

About 7 years ago, I could formulate a *first objection* to the principle of multiple embodiment. In fact, too often embodiments are *artificial* even though they appeared at face value to be concrete. Actually, concreteness has nothing to do with colors, motions, hate, or what not. If embodiments are not meaningful, then even though they are "concrete" they turn out to be as "remote" as the object itself. I then got interested in the notion of meaningfulness that, to me, appeared more basic than the multiple embodiment principle.

As for my second objection, it came only recently. In fact, we often forget that embodiments can be contradictory. Mathematicians often envisaged different enbodiments from an axiomatic perspective that bring an apparent unity. Situational factors often bring about undesired images, interpretations. The variety of personal distortions is not taken into account enough. I personally noted in my research that for the concepts of functions, directed numbers, rational numbers. . . a representation to become meaningful must be developed and at least mathematically be used by the students. “Intrinsic” familiarity with an embodiment is more important than the multiplicity of embodiment.

PHENOMENOLOGICAL PRIMITIVE
Ernst von Glasersfeld

“Phenomenological primitive” is an important notion and the way Andy uses it is very fertile. The moment you speak of phenomenological primitives, you are speaking developmentally. That I think is of paramount importance not only for education, because education deals with all these stages of development, but it is of paramount importance if you want to bring any clarity into the constructs you are using in analyzing because the context is something we build up. So the developmental view, I would say, is a very fertile one. Secondly, I think it is a powerful expression because it pins what everybody is trying to look at to the phenomenological, and phenomenological means belonging to someone's experiential world. That is important and the investigator must always keep in mind that whatever p-prim he attributes to the subject is his construction, not the subject's; it is the result of inference and therefore hypothetical; it can be tested only for compatibility and not for identity.

PIZZA PROBLEMS
Richard Lesh

There are those Pizza Problems where we asked: Jim ate a third of a pepperoni pizza yesterday and a fourth of a cheese pizza today. How much pizza did he eat altogether? Well, kids don't do very well on that problem. But if you give them the real version of the problem that is pull out a fourth of the pizza—Now Jim ate it. If he ate it, it's gone. So what do you have left?—you have a word left that reminds you of what was there yesterday: a fourth of a pizza. Next, you can say here is what Jim ate today: It's a third of a pizza. Now what is Jim dealing with? He is not dealing with a third or a fourth as symbols. He is dealing with a word and a pizza. So you've got a word–pizza problem. And if you look around out in the world there aren't that many symbol–symbol problems. They're usually symbol–pizza problems, word–symbol problems, word–pizza problems, all combinations of those. If you watch the solutions that kids go through in those problems, what you'll see is they're using often simultaneously (to characterize different pieces of the problem) different representation systems; that is, they may draw a picture to represent the static thing that is going on. Pictures do nicely to capture wholes. Once you have done that, how are you going to portray the transformation that's going to take place on those? Often kids skip to another representation system that allows them to deal with the manipulation. Often it is words. So each of the representational systems, words, written symbols, pictures have certain benefits and certain capabilities.

PLANNING
Andrea A. diSessa

Let me talk about planning for just a second. I think that the notion of planning is kind of a nice one. But, let me attack that one by saying that one really needs more than just the notion of planning. When planning, you sit back and you think about something in advance. You kind of get the feeling of what's going on, you have general categories of things that you're going to do, and you lay out a schematic advance. Then you do it. This has something to say about the way people solve problems. Here, I just state a prejudice. I'm quite confident at least in solving physics problems that planning of that sort has very little to do with at least the difference between experts and nonexperts. It's not a difference of skill at that level.

PLANNING, FORMAL NOTATION, AND IMAGISTIC PROCESSING
Andrea A. diSessa

So let me start with suitability of these ideas for us as theorists or experimentors in representation. My strongest agreement always comes in regions where he (Goldin) is talking about imagistic processing as being something important, as something that is very close to what we would like to talk about as meaningful understanding. But now let me go on to some problems that I have. Maybe the major problem is wondering whether these categories are in any sense "natural." Are they really distinctions of the world; are they cutting the world at its joints? Are these the right places to make cuts? They might be useful for some purposes but how widely useful are they? I immediately have some difficulties in that I look at these categories and I see them as functional distinctions. Planning, formal notational and imagistic, find largely functional rather than structural.

PRESENTATION VERSUS REPRESENTATION
Ernst von Glasersfeld

I would like to make a few brief remarks about a very naive conception of representation and its problems. Representation, I think, is an enormously ambiguous term. To show you some of the ambiguity, for instance, I could draw a circle here and tell you it is Paris; and I could draw a circle there and tell you it is London; and I could draw a line there and ask you what it is. Well, some of you might say it is the English Channel, and I'd say yes. It is a representation, there is no question about it. I could also draw something on the blackboard that looks like that and you would all say la Tour Eiffel, and it is also a representation. Third, in a more philosophical way, I could say that everyone of you has a representation of the world. Now, there are some very important differences between these three uses of representation. On the face of it, someone might say that the difference between 1 and 2 is that the second one is iconic. It gives you a picture of what it is supposed to represent. To that, I would say, so does the first one, it is only a question of degree. If by iconic or imagistic, you mean that the item that you call representation must conform, correspond to the original in some features that are considered crucial, then it is iconic. The distinction I make here is that I would call the first kind of representations iconic because both have some of the features of the original. When we come to the last one, a representation of the world, we are in a difficult situation, a situation that immediately evokes epistemology. Just to clarify my point of view, I don't believe that anyone of you has a picture of the world, in the iconic sense. The picture of the world is your own imaginative conceptualization of your experience. You are never in a position to say that the features of your picture of the world are in fact out there in the world. I repeat, this is an epistemological problem about which people think differently. If you think the way I do, then it is quite clear that you use "representation" in a totally different sense because there is no original. What you talk about is a construction and it can be a completely free presentation of yours. It is not re-presenting something else, it is creative, constructive, self-generated.

ROLES OF REPRESENTATION
Bernadette Janvier

When you talk about translation, I think children can do a translation to the extent that they understand each representation R1 and R2, but the problem with children is that usually they don't understand any representation. What I mean by comprehend is that when they see a representation they have a certain "feeling." If I want this to happen they need to construct this representation in a certain way; they should know the limits of this representation, what they can do with it, the advantages of one in relation to another. Students don't know what they can do with an equation or with a graph. They don't know what is carried by a representation.

THE CHILDREN ARE ABSENT
Maurice Bélanger

Perhaps this is the same idea as Bernadette's. I have felt during our deliberations that children are strangely absent from our discussion; this is similiar to the period of the 1960s when they were also absent in the new math and new science and there is a danger they will again be missing from the new representations. One of the things we need to remember is that children construct representations. We have much discussed representations constructed by adults. I suspect we could learn a great deal concerning representations for the learning of mathematics if from time to time we left the context of mathematics to examine how children construct representations in other knowledge domains. Everyone knows children love to draw. They draw people, for example. A number of researchers have analyzed such drawings, and we know that there is a typical "man-drawing" for children of 3, 4, and 6 years of age. This has been studied in detail. In contrast we are only beginning to study how children construct representations in other areas. I suspect that there are regularities that we don't know.

THE SYSTEM REPRESENTING ITSELF
Gerald A. Goldin

To sum up my comments, I think that the ideas in my chapter about language and problem solving do fit within this broad definition that Jim (Kaput) has offered us as long as we are careful to say that we cannot always specify in advance the components of the representational system. We cannot always specify the correspondence fully in advance. The system itself may be able to a greater or lesser degree represent that correspondence; the system itself may to a greater or lesser degree be capable of representing itself.

VERBAL SYNTACTIC IS IMAGISTIC
Andrea A. diSessa

I offer quick comment about verbal syntactic, as opposed to imagistic. I think that the distinction has some problems. One of my favorite examples is this nice sentence that everybody knows about that "Time flies like an arrow, and fruit flies like a banana" and if you wait for a minute then you reinterpret. "Oh I see, fruit flies, I see those little things that fly around, and they like, enjoy, a banana." You see an important shift in your understanding and it's a syntactic shift as well as a semantic shift. Those words change to different categories in your grammar of English language. I think that the statement that the syntactic level runs largely independent of the semantic levels is. . . well the categories are not totally independent, and in extreme cases, one might claim that there are an awful lot of connections between the two.

WE NEED A SYSTEM CAPABLE OF READING

Gerald A. Goldin

One important idea here is the idea of a system capable of reading the character set. If we have a representational system that these are characters, at what point does it become meaningful? Well the genetic code for example is only a code in a sense that there are molecules capable of interpreting the code by a chemical process. So to the extent that a representational system eventually gets understood by a person or read by a computer or interpreted by someone, then it always has a symbolic component to it, at least to the extent that the elements of the representational system symbolize some state of the machine or person that reads and interprets the representational system. So we need a system capable of reading.

WHAT ARE WE ATTENDING TO?

John H. Mason

If I say the word "two," then that abstraction resonates all sorts of experiences inside. I know each one of us has gone through the process of repeating a poem, and the recitation of that poem has come to something else so that we can speak about "two." That is a power you have. I want to look for the power that I want to activate. The ideas are the encoding but the activation of the power, the recognition I have of the power not just to count but to perform the act of counting, is in itself the idea of "two." For me a very difficult question is how can I tell right now to what each of you are attending? What aspect to what I'm talking about are you attending to? It might seem a very silly question but in a class of 35 students it is a very pertinent question. From the chapters I've read, most of us seem to be coming down with that question.

WHEN SYNTAX IS SEMANTICS

James J. Kaput

In the case of an algebraic system, as in the case of most practically used symbol systems, the syntax is not arbitrary. It is very closely tied to the semantics of the number system that, if you didn't look at the number system, you might regard as the rules for combining numbers, as again a syntax. Then a syntax at one place is semantics at another and so on. This is not an easy subject, otherwise it would have been solved long ago. The syntax of the algebraic representational system is guided in a necessary contingent way on a prior set of rules that evolved for numbers.

11 Pedagogical Considerations Concerning the Problem of Representation

Bernadette Dufour-Janvier, Nadine Bednarz,
Maurice Belanger
University of Quebec in Montreal

The purpose of this chapter is not to elaborate or to criticize what has been specifically said during this colloquium, but rather to contribute an analysis of some of our past and current research on the use that is made of representations in the teaching of mathematics and their effects.

We distinguish "internal representations" and "external representations." The first concerns more particularly mental images corresponding to internal formulations we construct of reality (we are here in the domain of the signified). The second refers to all external symbolic organizations (symbol, schema, diagrams, etc.) that have as their objective to represent externally a certain mathematical "reality" (we are here in the domain of the signifier). Even if this chapter is mainly concerned with external representations, we also examine how internal representations can be linked to external representations in learning.

Adults and children are daily confronted with a multiplicity of external representations. Mathematics teaching, school textbooks, and other teaching materials in mathematics submit children, even at a young age, to a wide variety of representations (Bednarz, 1984).

A number of questions may be raised concerning the use of these representations:

1. What are the motives for using external representations in mathematics teaching?
2. What are the expected outcomes that justify such a wide variety of representations?
3. Are these outcomes achieved in current teaching of mathematics?

4. To what extent is it possible that such representations are inaccessible to children and even detrimental?

5. Can the teaching of mathematics be organized in such a way that learning is articulated with the representations children develop themselves?

These are some of the questions that have been raised in the course of our research in this area. We examine each of these questions basing our discussion on examples and summaries of our research observations. Most of our examples are drawn from work with elementary school children because our past research has been conducted with this age group. At present we are also working with secondary-level students where we observe the same phenomena as those observed with presecondary students.

MOTIVES FOR USING EXTERNAL REPRESENTATIONS IN MATHEMATICS TEACHING

We can identify several reasons for having recourse to external representations. We rapidly examine a few of these.

Representations Are an Inherent Part of Mathematics. We are thinking here of conventional mathematics representations. One cannot pretend to have studied mathematics without having apprehended them. Certain representations are so closely associated to a concept that it is hard to see how the concept can be conceived without them.

Examples: functions and cartesian graphics
combinatorial analysis and hierarchial trees

For the mathematician, these representations are tools for treating these concepts. One hopes that the learner will perceive these representations as mathematical tools that he will be able to use appropriately in a given situation.

Representations Are Multiple Concretizations of a Concept. For example, several different representations may encompass the same concept or the same mathematical structure (Bednarz & Bélanger, 1983). In presenting these to the learner one hopes he will be able to grasp the common properties of these diverse representations and ultimately "extract" the intended structure.

Representations Are Used Locally to Mitigate Certain Difficulties. Mathematics textbooks and mathematics teachers make considerable use of such representations on a variety of occasions: (1) Once a task is given to a child, several representations are also furnished hoping that he will be able to find among these representations one that will enable him to accomplish the task (example, the child has to solve an equation and the author presents a number line, a set diagram and a pictoral drawing. See Bednarz, 1984); (2) in the course of the learning of a concept sporadic recourse is made to representations on which the child may lean to give meaning to what is being studied; (3) one wants to draw attention momentarily to a difficulty, or to an object or relation that he wants to put forth.

The Representations Are Intended to Make Mathematics More Attractive and Interesting. Authors use representations quite extensively in recent textbooks to embellish or ornament the presentation of mathematics to motivate the child or to present analogies to the real world.

Even though this is here only a partial inventory, we realize that there are numerous forms of representations to which a child is submitted, and furthermore that the objectives of using representations in mathematics teaching are quite diverse. We note also that certain of these representations have as a primary concern to be as accessable as possible to children, whereas others have as a primary concern the mathematical object itself.

Is this great diversity in form and intention necessary and useful?

EXPECTED OUTCOMES OF THE USE OF REPRESENTATION IN THE LEARNING OF MATHEMATICS

Even if we do not have major doubts concerning the necessity of introducing conventional mathematical representations that are inherently part of mathematics, we need nevertheless to interrogate ourselves about their utilization and contribution in the teaching of mathematics. We expect that the learner should perceive these conventional representations as mathematical tools. We also expect that he should be able, in a mathematical problem situation, to reject one representation to select another all the while knowing why he makes this choice. The representation rejected is not improper in itself but simply less effective in the given context. The learner should also be able to pass from one representation to another. All these expectations suppose that the learner has "grasped" the representations; that he knows the possibilities, the limits, and the effectiveness of each. For example, given an algebraic problem to solve, the equation and the graph do not give access to the same information and possibilities. We need to be able to choose the appropriate representation depending on the task, knowing why we make this choice.

As concerns multiple representations of the same concept we further expect that the learner will be able to grasp the common properties of these diverse materials and will succeed in constructing the concept. He should also be able to reinvest the knowledges acquired, in contexts embedding different aspects of the same concept.

Let us now turn our attention to the case where a task is submitted to the child accompanied by one or several representations. Knowing that he cannot succeed in the task without a representation, we expect that by intervening with a representation, the child will be led to solve the problem using the suggested representation. Another expectation, when using several representations for the same task, is that the child develops positive attitudes such as trying to find a representation that could help him when faced with a problem, as well as trying to approach the problem from different points of view.

Concerning the representations used to attribute meaning to a symbolism we expect that as soon as a child has difficulties working with this same symbolism, he will have spontaneous recourse to one of those representations.

We have just surveyed some of the behaviors we expect children should exhibit if teaching makes extensive use of representations. If these behaviors do not occur we can then raise serious doubts on the appropriateness of these diverse representations or the way they were used.

ACHIEVEMENT OF THE EXPECTED OUTCOMES IN THE CURRENT TEACHING OF MATHEMATICS.

Several questions are now raised in relation to the expectations stated previously. In an attempt to answer these questions we summarize observations we have made during the period of our research and present a few examples.

Are conventional mathematics representations perceived as tools for solving problem situations?

Let us take the example of Cartesian graphics. A group of high school students are given the following equation to solve: $\sqrt{x} = -x + 6$.

They were in the process of learning a new algorithm to solve irrational equations; an algorithm that the teacher had listed in five steps. The students passed through the first four steps, which were listed on one side of their sheet of paper, without anticipating the answer, or questioning or reflecting in a manner that indicated that they attributed any meaning to what they were doing. After completing the fourth step of the algorithm they had obtained two solutions: $x = 4$ and $x = 9$. One of us, circulating in the class, told a group of students that the solution was $x = 4$. They protested that they had correctly gone through the four steps without making any mistakes and had found two solutions. To counterargue we asked them to trace the graph of $y = \sqrt{x}$ and the graph of $y = -x + 6$. They easily managed to do this. But they again protested, "that's not what the teacher told us to do.'' Even with the two graphs they had drawn directly in front of them, where there was a single intersection point manifestily not at 9, they continued to insist that there were two solutions. The classroom teacher who noted the same difficulties said, ''You've forgotten to turn the page indicating the fifth step of the algorithm!!,'' which was to verify the solution(s) by substitution in the original equation. This is what finally convinced them that $x = 4$ was the correct and only solution!!!

This is only one example, but unfortunately not an isolated one. Even if the children have studied the mathematical representations and can produce and use them on demand, they do not have the attitude of turning to these as tools to help them solve problems. These representations are perceived as mathematical objects in themselves, but not at all as means towards a solution of a problem.

How do concrete multiple representations used in current teaching contribute to the child's construction of the concept?

We have observed that all too often ''multiple representations'' are merely simplistic correspondences to the concept involved. In this case the elementary school children we saw were hopelessly lost when placed in shifted contexts involving the same concept. We were thus led to have serious doubts in such cases about the effectiveness of multiple representations as contributing to the child's understanding. More often it is a game of putting into correspondence one representation with another. What is constructed in such cases are syntactical rules of correspondance rather than constructing the concept itself via its representation.

Example: Children are asked, in individual interviews (Bednarz & Janvier, 1984), to represent on an abacus the number 3152. They did this by using the rules of correspondance. We then wrote 128 on a piece of paper and asked them to subtract that number on the abacus. Almost all children did the subtraction on the abacus trying to put the rules for paper–pencil subtraction in correspondance with the abacus. All sorts of errors were committed leading to different answers. We then gave them (on the same paper where we had written 128) the subtraction 3152 − 128. We then asked ''can you tell me what the answer will be?'' The children's responses inform us whether they recognize that the task is the same as in the previous one. Only one child said that he had already done this problem; the others said they had to calculate. All sorts of errors were committed in their calculation and even when the result was different from the calculation previously done on the abacus, this did not at all upset them. Some children redid the subtraction (using vertical calculation) and obtained yet a third answer. They considered it natural to obtain three different answers because they had done the calculation three different ways. In fact they do not see the three representations as embodying the same situation.

To what point can a child, who is experiencing difficulty in solving a problem, select a representation that will bring him to resolve this same problem?

Although we do not answer this question directly, it raises several other questions that may be addressed on the basis of our research findings.

Does a Child Really "Select" a Representation? Among several representations presented to him, does he know which one to retain, which is the most appropriate to accomplish the task?

The young children we have seen in interviews, who cannot solve the problem posed, do not really select a representation. More often they retain the one that is most familiar to them or make remarks such as, "With that one I can count better," or else retain the representation where numbers in the problems are made more explicit. Obviously, they often do not retain the most appropriate representation and may even use the representation selected completely forgetting the original problem posed.

Example: We present a problem to a child involving the notion of distance travelled. The problem is accompanied by a number line, a set-diagram, and a drawing of a "function-machine." The child attempts the problem directly on the set-diagram without necessarily arriving at the correct solution. In such a case he has retained the representation that is most familiar to him without realizing that the set-diagram is not very appropriate as an aid to solving problems involving distances.

Does the Child See the Same Task in Each of the Representations Given? The explanations given to us by a number of children reveal that they see as many different problems as there are representations. They have a tendency to say that two representations concern the same problem only if the same numbers are evident in both representations (Bednarz & Janvier, 1984).

Is the Child Convinced That Regardless of the Particular Representation He Uses as an Aid to Solve a Problem He Will Necessarily Arrive at the Same Result? Example: A child resolves a problem using a representation. We then show him the same problem resolved by someone else using a different representation. When we deliberately show him the answer of this other child (the answer happens to be incorrect); a number of children are not at all disturbed and find this quite natural because in their view the first problem was done one way and the second done in another way.

This is an illustration where the child expresses the conception that a problem done in different ways may lead to different answers.

It is thus far from evident that the child sees all these representations accompaning a task, as different ways for tackling the same situation.

Finally, a few comments on another expected outcome previously stated.

How Do Children Develop the Attitude of Having Recourse to Representations in Case They Encounter Difficulties? This attitude is fostered if they acquire the conviction that these representations can be a help in unscrambling a problem situation.

Example: 3276 divided by 2 (posed to third-grade children in a numeration test). In the group (having received traditional teaching using a variety of representations), the children blocked, having no other recourse possible other than turning towards certain rules. In the second group (for whom the teaching was based on representations developed by the child (Bednarz & Janvier, 1985), the children spontaneously translated the given facts into another representation with which they were more at ease to operate and resolve the problem. This example illustrates that a translation is possible to the extent that the child has well "grasped" each of the representations.

We have a few additional comments to make before terminating this section on representations, intended to be motivating or embellishing.

Quite often when the intention is to represent some real-world situation, the analogies presented to the child do not lead to the desired connections. The child glances at the representation and fails to see any relationship with the mathematics he is learning. We do not need to elaborate this point much further to be convinced that it is not simple references or superficial artifices that will enable us to achieve the intended objectives (Bednarz et al., 1981). Focus needs to be placed more on the way that mathematics should be presented or taught.

In the plethora of representations used in teaching materials we have observed that children do not even bother to look at them.

Example: We present to the children a sheet of paper on which there is an equation to solve. Above the equation a set-diagram is given of the same problem. Most of the children go directly to the written equation using their own representations (for example, their fingers or tick-marks) that often lead to the wrong answer. We return to the sheet and ask each child "What was the problem you had to do? Was there anything else on the sheet?" To the first question the responses are all of the same type: "We were searching for a number, we had an equal, we had to find what to put in the space." Certain children could not answer the second question. Others, after reflection, said there was something else, maybe a diagram, but they were not sure. Still others remembered having seen a set-diagram and were ill at ease that they had forgotten to do this "other" problem.

The child comes to the point where he no longer sees what is important in what is presented to him. Either he starts thinking that everything presented is pertinent, or at the other extreme, that all is accessory and, in a sense, nothing is relevant. Either case is equally bad. In the first situation the child faces a host of difficulties and problems because he tries to find a logic that incorporates all the elements. In the second case the child is left with not knowing what to do and requires constant explicit directions.

The picture just outlined is not very encouraging. We note in fact that in current teaching the use of representations frequently does not lead to the attainment of the desired objectives, and that in certain cases their contribution is nil. Serious doubts need to be raised on the use that is made of representations especially when, as we have observed, certain representations lead more to difficulties rather than functioning as aids to learning.

If we want representations to be really useful in learning mathematical concepts, a number of questions will have to be considered: Which representations should be retained? Are there representations that are more appropriate than others for developing a concept? How should the representations be used? In which contexts? What are the difficulties and the children's conceptions that need to be taken into account when a representation is used? Are there representations that are more appropriate to the level of development of the child and to where he is in regard to the learning of mathematics?

We now discuss only the last question because we think this to be of critical importance and by doing so touch some aspects of the other questions.

ACCESSIBILITY OF REPRESENTATIONS

Most people who encounter a child with difficulty in resolving a problem focus on the lack of understanding of the mathematical concept involved, rather than on the representations that are utilized. Too few people seem to be preoccupied with this dimension. As previously discussed, representations will be useful to the child to the extent that they have been "grasped" by him. But how can a child be brought to appropriate a representation that is initially inaccessable to him?

The premature introduction of representations can sometimes explain their uselessness, and furthermore they may even have negative effects on learning. Here are a few examples.

Having verified that a group of first-grade children can put in proper order 3 or 4 positive integers, they are then asked to employ the sign ">" to write a string of numbers in decreasing order. This task is virtually impossible for certain children. At this age level the use of signs (such as ">") is inappropriate, as can be further illustrated by their difficulties with the inversions in writing letters and digits.

The use of two-dimensional tables also causes numerous problems to young children. Intended as representations to help children organize information or data, in fact children confront the difficulty of knowing which cell to fill out, and then how to use the table to bring out properties. The child is so preoccupied with the mastery of the instrument that this works against the mathematical knowledges and skills that were intended to be developed.

As a consequence, in these two examples, the teaching of a symbol (<) or a table becomes an end in itself and not means to help the child to organize or have access to mathematical knowledge.

The premature use of a representation, as well its application in an inappropriate context, can lead children to develop misconceptions that will hinder them in later learning. Let us examine some of the consequences.

First, when children use the number line during the learning of positive integers, they develop the notion of the number line as a series of "stepping stones." Each step is conceived as a rock, and between two successive rocks there is a hole! This is very far from the concept of the density of the real numbers as illustrated by the number line. It is hardly surprising that at the secondary school so many students say that between two whole numbers there are no numbers, or at most one. Nor should there be much surprise that they also have great difficulty placing a number if they cannot associate it with the gradation already given on the line. Or again if in their reading of graphs they focus only on the values given by the gradations.

Secondly, many children see the numbers on the number line as though they were "number-panel-signs" placed along a road. They don't see any necessity for placing these billboards at equal distances, saying only "well, usually they are equally spaced." Here this is far from the measure concept, also illustrated by the number line.

Still a third hindrance in conjunction with the number line, and also in many other mathematical representations, is the use of arrows. This graphical code alone can have different meanings: displacement, transformation, pointer, cause–effect relation, etc. However, the meanings that young children usually give to the arrow are a pointer or a line where the two extremities indicate two states or two positions. If the arrows are curved over a number line, then there are some children who interpret these as strings delimiting sets.

For them, the following diagram represents $1 + 5 = 12$. Note that 1 and 5 are included "inside" strings.

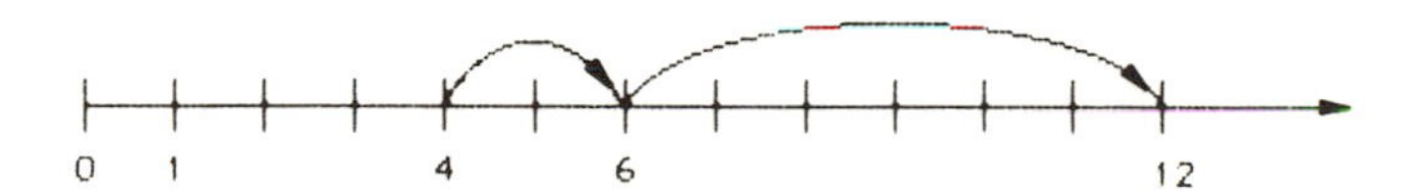

The number line is often used in an attempt to give a sense to equations. However, children develop numerous misconceptions of the number line. It is then hardly astonishing that children have such a poor comprehension of the equivalence relation (Belanger & Erlwanger, 1983).

The imposition of external representations that are too distant from the child results in having the child react negatively or causes him difficulties. If one wants to use an external representation in teaching, he needs to take into consideration that it should be as close as possible to childrens' internal representations. However, one frequently imposes external representations without realizing the large gap that can exist between those used and the one envisaged by the child of the problem situation. Young children want to find the characteristics in a representation that they perceive as essential to the situation studied. At a moment in their learning they thus respond negatively to a representation where "we don't see what's going on": for example, in the case where they have to carry out a sharing by doing the conventional division algorithm or by using canonical decomposition of a number.

Example: In a third-grade class the children had seen and done several activities of sharing. They had used such materials as groupings of flowers each with 10 petals, and bouquets of 10 flowers (Bednarz & Janvier, 1985). The children were suppose to share equally on paper a collection of flowers among a number of persons. In their initial representations the children insisted on drawing each of the persons who were to receive flowers. Even if they finally abandoned this very descriptive representation, they still remained attached to it for a long period of time.

The evolution is very long and needs to be provoked in such a way that they come to see that what is essential in the representation is not the act of distributing to all the persons, but rather what each person receives. However, the two procedures most commonly taught for divisions do put the emphasis on the latter (i.e., what each person receives) but in no way puts in evidence the act of sharing or unmaking groupings. We can then anticipate from this example the difficulties provoked when what is taught is not in conformity, or too far away from the representation that the child formulates.

The use of representations that are just as abstract for the child as that which is studied brings the child to manipulate rules and symbols that are meaningless to him. For example, to teach numeration the abacus is used without taking into consideration the age of the child and where he is in regard to the acquisition of a concept. Diverse operations can be illustrated with this apparatus. The young child who uses the abacus to carry out his operations applies rules that are just as conventional as when he operates with numbers, because for him the grouping rule is just as conventional and not apparent.

In analogous cases to those just presented the child is forced to "learn" the representation that is submitted to him: the rules of usage, the conventions, the symbols, and the language linked to the representation. This "learning" of a representation is often done to the detriment of the notion that was to be learned.

The analysis we have presented made us realize the negative consequences that can be caused by the use of representations prematurely or in an inappropriate context. In fact this leads the child to develop erroneous conceptions that will subsequently become obstacles to learning. The use of such nonaccessible representations encourages a play on symbols, puts the emphasis on the syntactical manipulation of symbols without reference to the meaning. The signified is absent! Mathematics is reduced to a formal language.

As a consequence one must be very careful in the use of a representation. It must be kept in mind that the characteristics perceived of the situation by the child may change, and that the motives of the child for using a representation can be very far from the intentions of the person proposing them. These two points of view need to be reconciled, but we have realized that this correspondance, which is necessary for learning mathematics, is difficult to realize.

Could this reconciliation of points of view be facilitated if we had recourse to the representations developed by the children? They, in fact, construct a variety of representations in domains other than mathematics. They draw, mime, and verbalize their actions. There is a need to examine how children use objects, how they act, and the representations that they construct. A study of these might furnish us suggestions for formulating alternative conciliatory representations for the learning of mathematical concepts.

CONTRIBUTIONS OF CHILDREN'S REPRESENTATIONS TO THEIR LEARNING

In pedagogical methods currently used in schools, external representations are imposed from the outside, and there seems to be very little opportunity or encouragement for children to construct and exploit their own mathematical representations. The only moments when children are prompted to construct representations are occasionally in the case of word problems. Appeals are made to produce a drawing, "try to make a drawing," or else, "draw and you will understand." We wanted to know if these appeals to drawing, that are not integrated into regular practice, bear fruit.

Can we bring a child to solve a problem by coercing him to produce a drawing?

We have given a series of complex problems to two populations of children (group A and group B) from first to sixth grade. The directions given to group A were to solve the problems in such a way that we could "see" what they did. We told children in group B about another group of children (group A) who had been unable to do the problems in spite our attempts to furnish help, and that we came to see them so that they could furnish us means, or tools, to help the children in group A understand the problems. These children proposed to us means ranging from the suggestion to rewrite the problem to actually giving the answer. When the children of group B had finished this task, we then asked them to solve the problems. Almost all the children in group A gave us the answer and very simplified calculations, whereas children in group B produced drawings. However, there was no overall difference in the level of solving the problems correctly.

The drawings produced by the children in group B were for the most part simple contextual correspondances that did not in any way make clearer the facts or relations contained in the problems. It would seem then that the children could not produce, on the spot, effective representations to help explore or solve the problems. This study gives results that are not very encouraging. They reveal that the appeals to drawing do not produce the understanding we might have expected. For this understanding to occur we should have the objective in teaching mathematics of bringing children to disengage the particularities of a problem situation by having recourse to a drawing.

In our research on the learning of numeration (Bednarz & Janvier, 1984, 1985), this was one of the objectives we had in mind. For this purpose we elaborated a strategy in which we brought children to develop codes, and symbols, and to refine their representations. At the beginning, descriptive as these representations were, they gradually evolved to more formal and effective representations. Through this process, the children really came to perceive the role of representations that is to put into evidence the particularities of a situation, all the while enabling the modeling of actions and procedures (Bednarz & Janvier, 1985).

Another dimension, and not the least that should be considered, is the role of language.

How does language intervene in the process of constructing representations?

We observed in our research that the child can distance himself from a problem situation to the degree that he can refer to a previously developed discourse on the actions and procedures that he performed or observed in a real situation representing the problem. The first external representations produced by the children were intimately tied to this discourse. The children wanted to find on paper all that was said, or done, by themselves or others. In the initial stages the children attempted for a long time to represent everything by writing and drawing, even if they realized that the representation produced was ponderous to operate with. Their conception of a good representation was that it should contain everything presented and no information should be lost.

When imposing external representation, in current teaching, insufficient importance is given to the role of language. Also there is a lack of awareness of what children consider as a good representation. This helps us understand the discordance, between the point of view of the teacher and the child when external representations are not accessable to children.

From this perspective of having children develop their representations, one may raise the following question: What should a child dispose to be able to construct a representation?

One might think that it is the skill to draw that is the critical factor. However, the facts we have noted during the 3 years of working with the same group of children are revealing. In the class we worked with, certain children were known as skillful drawers (two were considered excellent). But in the course of the 3 years only one of these children, excellent at drawing, consistently had recourse to producing drawings when presented with a problem. He was one of the children who often had difficulties using their representations because they contained too much unhelpful and distracting information.

All the children wrote rather than drew, even if in the first, second, and third grade children are generally not more skillful in writing than in drawing. This insistance to write can be explained by their primary preoccupation that representations should capture actions, transformations, and relations. To represent these they need to have available appropriate graphic codes and symbols. Because they lack these codes and symbols they instead make extensive use of written language. It is not the ability to draw as such that is of importance, but rather the capacity to develop graphic codes and symbols that will effectively represent a given situation and suggest actions.

Can we expect that by developing their external representations children will construct their knowledge? Our experimentation of such a strategy has convinced us that the external representations of children can be informative of how they perceive a problem situation and thereby provide guidelines for designing interventions. The then desired reconciliation of points of view between teacher and child is facilitated and we believe is realized. However, such teaching is very different from traditional teaching because it calls for a change of attitude. It presupposes a better knowledge of external representations of the children, of the conceptions attached to a concept corresponding to external representations of a certain reality, of the obstacles encountered by the child of a representation (that he developed or has been proposed to him), and a dynamic knowledge of these representations—how they evolve in close relation to the signified.

CONCLUSIONS

Regardless of the specific objectives envisaged concerning the use of representations, the overriding purpose should be that they are helpful to the child so that he learns to use them effectively. In mathematics a host of conventional representations already exist; concerning these there are a number of important research questions needing investigation: Which representations are more interesting from the point of view of the child?; which are in closer accordance to his development?; and what are the contexts of their appropriate use?

The child constructs representations in many domains in order to discover regularities that would enlighten us on graphic codes and symbols for formulating alternative representations. It would also be interesting to do a developmental project that would explain how the ability of a person to represent in a variety of domains develops, and how this ability interacts with natural language.

We can help a child to construct his knowledges by a learning strategy that is articulated with the representations developed by him. We have experimented with such a strategy in the course of our research on numeration, and it would be desirable to undertake additional studies bearing on other mathematical concepts. These studies would enable us to examine childrens' intermediate representations, their conceptions, and the obstacles that need to be faced when external representations are imposed for developing a concept.

The area of the representation of dynamic situations (involving action, transformations, movements) is particularly intriguing and corresponds to domains where children encounter learning difficulties (Bednarz, Dufour-Janvier, & Poirier, 1983). We have observed that representations in current teaching materials, intended to evoke dynamism in a situation, are not decoded as such by the children. In our current research we are investigating how it is possible to construct representations of dynamic situations and to what point these representations can be a support to learning these concepts. In particular, what are the graphic codes used by children to represent situations where a certain dynamism is at play?

ACKNOWLEDGMENTS

We thank our research associates, in particular Louise Poirier and Diane Biron, for helpful comments on this chapter.

REFERENCES

Bednarz, N. (1984, May). Problèmes de représentation en arithmétique au primaire. Rapport de recherche, *Cahiers du CIRADE.*

Bednarz, N., & Bélanger, M. (1983, July. Problems of representation of an operation in elementary school arithmetic. *Proceedings of the Seventh International Conference of PME, Israel.*

Bednarz, N., Bélanger, M., Janvier, C., Janvier, B., & Boileau, A. (1981). Un exemple de "sur-symbolisation" dans l'Enseignement des Mathématiques. *Instantanés Mathématiques.*

Bednarz, N., Dufour-Janvier, B., & Poirier, L. (1983, October). Problèmes de représentation d'une transformation en arithmétique chez les enfants du primaire. *Bulletin AMQ.*

Bednarz, N., & Janvier, B. (1984, November). La numération: Les difficultés suscitées par son apprentissage—Une strategie didactique cherchant á favoriser une meilleur comprehension (1ere partie). *Revue Grand N,* IREM de Grenoble.

Bednarz, N., & Janvier, B. (1985). La numération: Les difficultés suscitées par son apprentissage—Une strategie didactique cherchant á favoriser une meilleur comprehension (2ime partie). *Revue Grand N,* IREM de Grenoble.

Bélanger, M., & Erlwanger, S. (1983). Strategies used by first to fourth grade children to transform statements containing an equal sign. *Proceedings of the Fifth Annual Meeting of the North American Chapter of IGPME,* Montreal.

III
FIRST STEPS TO THE SOLUTION

12 Cognitive Representational Systems for Mathematical Problem Solving[1]

Gerald A. Goldin
Rutgers University

INTRODUCTION

This chapter presents some perspectives on the general concept of *representation,* and the more specific notion of *cognitive representation,* which bear on the framework for describing competence in mathematical problem solving proposed in my earlier work (Goldin, 1982, 1983, 1985a).

The chapter is organized as follows. First the question of defining a *representational system* is addressed, and several illustrative examples are discussed. Attention is given to the role played by ambiguity. *Symbolization* is then considered as a relationship between two representational systems and, among other issues, its use to resolve ambiguities and the interpretation of "context-dependent" symbolization are examined. These constructs allow increased precision in the analysis of several important philosophical questions that pertain to the issue of cognitive representation—e.g., the ontological question of what cognitive representations "really are," and the epistemological question of what "reality" (if any) cognitive representational systems actually represent. The paper continues with further discussion of five kinds of cognitive representational systems comprising my present model for mathematical problem-solving competence, and the degree to which it makes sense to call these systems *languages.* It concludes with a statement of my position on the most desirable goals of mathematics education, and a speculative comment about possibilities for the future evolution of mathematics.

[1]An earlier version of this chapter was presented at the American Educational Research Association Annual Meeting in Chicago, Illinois (April 1985).

Credit for many of the ideas discussed belongs to others. For example, distinctions between structure and function, syntax and semantics, and logical languages and natural languages, as well as the implications of viewing mathematics as a collection of wholly formal axiom systems apart from the world of empirical observation—these well-known ideas have their origins in the pioneering work of mathematicians, philosophers, and linguists, whom I have tried to mention in the bibliography. However I have not traced the history of their views in the published literature, and thus make no claim of bibliographic adequacy in this regard. Instead I have tried to be both eclectic and critical, drawing on various concepts from these fields as needed and modifying them freely in order to address the issue of cognitive representation in problem solving with greater precision (Chomsky, 1965; deSaussure, 1959; Hofstadter, 1979; Korzybski, 1958; Kline, 1980; Nagel and Newman, 1958; Piaget, 1969, 1970; Quine, 1960; Russell, 1919, 1940; Skinner, 1953, 1957, 1966)

In addition, I owe an intellectual debt to the other participants in the conference organized by Claude Janvier and sponsored by the University of Quebec at Montreal in June 1984. The reader is referred to their papers and the accompanying discussions in the present book.

REPRESENTATIONAL SYSTEMS: DEFINITION AND DISCUSSION

Kaput (1983, 1985), following Palmer (1977) and others, focuses attention on describing the following components in specifying a *representation*:

(a) a represented world;
(b) a representing world;
(c) aspects of the former which are represented;
(d) aspects of the latter which do the representing; and
(e) the correspondence between them.

In asserting the need for these five features, Kaput seeks a general framework including not only perceptual and cognitive representations, but explanatory models in science, mathematical models of various kinds, and external symbolic representations. In discussions during the Montreal conference, both von Glasersfeld and Mason questioned the use of the term *representation* in perceptual and cognitive contexts, arguing from an epistemological standpoint that the act of mental construction must be taken as *primary*. In their view, the inner world of experience should therefore be considered a *presentation,* rather than a representation of some other external reality.

Because I am principally concerned with cognitive representations, which many researchers consider essential to constructing a meaningful theory of natural-language understanding as well as problem solving (e.g. Greeno, 1983; Johnson-Laird, 1982; Lesh, Landau, and Hamilton, 1983; Palmer, 1977), it would seem worthwhile to begin by defining some concepts related to the term representation with greater precision, with the goal of being able to respond definitively to this comment.

Let me begin by discussing what might reasonably be called a *representational system,* generalizing the term used by Lesh, Landau, and Hamilton (1983, p. 270). This construct is intended to include the notion of a "symbol scheme" discussed in Chapter 14 by Kaput, spelling out in more detail the representing world mentioned above.

A representational system shall consist first of a collection of elements called (interchangeably) *characters* or *signs.* These elements are primitive in the sense that they do not at this point stand for or symbolize anything else. They may be drawn from a well-defined, finite, discrete set (such as letters of the Roman alphabet or characters in the sentential calculus), or from an infinite set (such as closed curves in a bounded region of the Euclidean plane). Alternatively, the elements may be drawn from an ill-defined class, such as "real-life objects," or "drawings on a sheet of paper." If a character is considered to be an equivalence class of *inscriptions,* a possibility noted by Kaput during the conference, then it may be that the equivalence relation involved is itself poorly defined. In entertaining at the outset this possibility of an ill-defined class of primitive signs, we encounter the first *ambiguity* which can occur in the discussion of representational systems. Later we shall discuss ways of resolving such ambiguities. For the present, we simply note its presence without becoming too uncomfortable; after all, an ambiguity subject to later resolution does not necessarily introduce any contradictions into our definition.

At the risk of redundancy, let me emphasize here how extremely restrictive is the assertion that the characters do not symbolize other entities. Thus, if the characters are specified to be the printed Arabic numerals 0, 1, 2, 3, 4, 5, 6, 7, 8, 9 (by which we understand not merely the single inscription of these numerals on this page, but a conventional equivalence class of such inscriptions), then it is not correct (*within* the representational system) to say that the numerals stand for the cardinalities of certain sets, or even to say that the numerals have pronunciations or spoken names. Later, cardinalities of sets, spoken names, and so forth, can be introduced separately as characters in other representational systems, and correspondences between the appropriate elements of different systems can be asserted.

A very elementary device, such as an on-off switch, may for practical purposes require no further specification as a representational system than the mere listing of its signs. In general, however, it is convenient to let our definition of a representational system allow for additional structure. We shall see below that this is entirely a matter of convention: an alternative is to say that representational systems consist only of primitive signs and to introduce additional structure entirely through correspondences or relations between such representational systems. The latter approach, however, would quickly lead to an unmanageably large number of distinct representational systems. Without claiming to offer an exhaustive characterization of all the possibilities, let us mention some important ways in which the additional structure can be introduced into a representational system.

First, there may be specified a set of conditions which describe *permitted configurations of characters* in a representational system. Such conditions might consist simply of an *inventory* of allowed configurations—for example, a vocabulary list of words written in letters of the Roman alphabet. Alternatively, the conditions might be *rules* describing how permitted configurations are generated, or constraints that they must satisfy—for example, the rules for writing well-formed formulas in the first-order predicate calculus. Now, just as the class of characters or signs of a representational system can be ill-defined, the conditions for forming permitted configurations can in the general case be ambiguous; that is, implicit, fuzzy, only partially specified, or having unstated exceptions. Here we have a second ambiguity subject to later resolution.

In the historical development of mathematics, completion and rigorization of the systems of rules governing formal notations (i.e., elimination of ambiguities) became an important goal, leading in the 19th and 20th centuries to fundamental reconceptualizations of the nature of mathematical reasoning (Kline, 1980). In many situations outside of mathematics, however, including the grammar of natural language (see below), little insight appears to be gained by trying to complete or make rigorous the conditions defining permitted configurations in a representational system. Instead, this process leads to an unwieldy collection of *ad hoc* rules. Thus, for the present we will accept the maintenance of ambiguities as the lesser evil.

Suppose now that we have specified, albeit ambiguously, a collection of characters and a collection of permitted configurations of characters. The following are some examples of what might be called *higher-level structure* that can next be introduced into the definition of a representational system:

(a) rules for forming *configurations of configurations* (higher-level configurations);
(b) *networks* or graphs of permitted transitions from one configuration (or collection of configurations) to another;
(c) *relations* on the collection of configurations (e.g., partial or total orderings);
(d) rules for assigning *values* to configurations, (e.g. truth values, goal values, etc.);
(e) *operations* on the collection of configurations; or
(f) combinations of these and/or other kinds of structure.

These examples are, of course, not intended as an exhaustive system of classification, but merely to be suggestive of the kinds of higher-order structures that frequently occur. Again, we must allow for the possibility of ambiguity in the introduction of any of the above structures.

Let us now consider some illustrative examples.

Example 1. In logic and mathematics, certain allowed configurations of characters (well-formed formulas) are designated initially as axioms (cf. (d) above). Rules of valid procedure are then specified (transitions from collections of configurations to new ones, cf. (b) above), whence other configurations can be obtained ("proved") as theorems, starting from the axioms and using the given procedural rules. It is again important to stress the fact that neither the axioms nor the theorems are being regarded here as "truths" about physical objects, mathematical entities, or anything else. Truth values can be assigned, but they are simply an arbitrary way of labeling configurations (cf. (d)). Here the goal is, if possible, to eliminate any ambiguities in specifying the representational system.

Example 2. In the case of a finite-state machine (such as a Turing machine, which reads signs on a tape), the characters of the representational system might be taken to be sets of values for variables that describe the state of the machine (e.g., Hunt, 1975, p. 33). Rules describing permissible states define the configurations of the system. Higher-level structure is provided by specifying the network of possible transitions from state to state (cf. (b) above). In this example, we have not yet specified the *contingencies*—that is, how the selection of the transition which actually occurs (from among all those which are possible in a given state) depends on the sign or signs read.

Example 3. A measuring device with a continuous set of possible readings, such as a thermometer, is also a representational system whose characters or signs may be taken to be the physical constituents of the instrument. The permitted configurations of these constituents form the set of possible states or "readings" of the device. Here, ambiguity may occur in establishing criteria for deciding when two readings are considered the same (error of measurement). Higher-level structure may, for example, consist in a linear ordering of the set of possible readings (cf. (c) above). As with the other examples, we have as yet said nothing about the actual use of the device to measure anything.

Example 4. In the analysis of task structure (and in artificial intelligence research generally), a useful tool is the *state-space representation* (Nilsson, 1971; Goldin, 1984a). Here a state is a permitted configuration of a problem (or game), whose signs may be mathematical characters, components of a physical device such as a chess set or a "Tower of Hanoi" puzzle, points and line segments as in "Pen the Pig," and so forth. Additional structure is provided by designating an initial state and describing one or more goal states (for each player, in the case of a game), and specifying the allowed moves or transitions from state to state [cf. (b) and (d) above]. However, the states do not at this point symbolize or represent anything beyond themselves. The state-space representational system does not itself include algorithms, strategies, or heuristic processes for actually making moves within it.

Example 5. This example might be called purely verbal, nonmeaningful printed English. Take the characters to be those on a typewriter keyboard, and the permitted configurations to be words in a vocabulary list, together with punctuation marks. Higher-level structure is provided by ''standard'' rules of English grammar for forming syntactically admissible sentences from configurations that are English words or punctuation marks, thus yielding configurations (cf. (a) above). Application of these grammatical rules requires that the words first be labeled by part of speech—as nouns, verbs (transitive or intransitive), adjectives, etc.; thus, part of speech might be considered a value assigned to each configuration (cf. (d) above), and there must be more than one value allowed for many words. We call the example ''nonmeaningful'' English because no *semantic* interpretation of the words as standing for ''real-life objects,'' ''the world of experience,'' or anything else for that matter, is implied.

Now it was stated above that instead of regarding a representational system as a single system having one or more of the kinds of structure mentioned, one could introduce the structure as a (possibly ambiguous) correspondence or relation between representational systems consisting only of primitive signs. In Example 1, for instance, one could take the set of characters in the sentential calculus to be one representational system, the set of well-formed formulas (themselves regarded as primitive) to be another system, and define relations between the two systems which express the fact that formulas are ''spelled'' by characters. One would then take the set of truth values to be a third representational system, and rules of valid inference to be a fourth. Likewise, in Example 5, one might take the set of typewriter characters, the collection of English words, the set of parts of speech, and the collection of all grammatical sentences, to be four separate representational systems. One could then consider the spellings of words, the assignments of parts of speech to words, the parsings of sentences, the verbal composition of sentences, and so forth, all as relations between characters from one system and characters from another. This is the sense in which the decision to say that we have one representational system or many is a matter of convention.

In the next section, we discuss symbolization and its use in resolving ambiguities that occur in the definition of representational systems.

SYMBOLIZATION

Let us use the word *symbolization* to refer to the correspondence or relationship asserted by Kaput between two representational systems—one representational system being the "represented world," the other being the "representing world." The aspects of these worlds that are represented and that do the representing will be the characters configurations, or higher-level, structures in the two representational systems. Kaput has observed that the designation of one system as represented and the other as representing is arbitrary, and can be reversed as needed.

Just as the signs and configurations of a given representational system can be ill-defined or only partially specified, so can the correspondence between two representational systems be fuzzy. A mapping in mathematics may permit one system to represent another in a very precise way, as when an abstract mathematical group is represented by linear operators in a vector space. On the other hand, words in a purely verbal representational system can represent objects in a nonverbal representational system, yet the relationship may defy precise specification.

Now that we have introduced the notion of symbolization, it should be observed that there may be choices that fall between the two conventions discussed in the first section of this chapter—that is, regarding a system with higher-level structure as a *single* representational system, or regarding it as many representational systems comprised only of primitive signs. One could equally well describe the structure of such a system by partitionng it into simpler representational systems, and taking one of them to symbolize the other. For instance, we might separate "nonmeaningful printed English" (Example 5 above) into two representational systems—English spelling, whose characters are letters and whose configurations are spelled words, and English grammar, whose characters are punctuation marks and words tagged by part of speech and whose configurations are grammatically permissible punctuated sentences. The former symbolizes the latter by spelling its words. Thus there is great flexibility as to the level of detail included within a single representational system.

Now we return to the issue of resolving the ambiguities permitted in the definition of representational systems and symbolization. As an example, let us consider the amusing pair of English sentences cited by diSessa during the Montreal conference:

(1) Time flies like an arrow.
(2) Fruit flies like a banana.

Because the words "time" and "fruit" can be tagged either as nouns or as parts of noun phrases, the word "flies" either as a verb or as a plural noun, and the word "like" either as a preposition or as a verb, the grammatical constructions of these sentences are not uniquely determined *within* the purely verbal representational system under discussion. In order to select the semantically sensible construction, reference must be made to another, possibly nonverbal, representational system—for example, a system for visualizing real-life situations—within which it is feasible to conclude that one of the alternative grammatical constructions for each sentence "makes good sense semantically in the present context," while the other does not. Without reference to such a second representational system, it is difficult to avoid including the nonsensical as well as the sensible constructions of the above sentences—as well as sentences such as "Time likes a banana" that are grammatically correct but have no sensible interpretation in most contexts.[2]

This example illustrates what seems to me to be a widely applicable general principle in the theory of representational systems:

> *Ambiguities* within *a representational system, either in the definition of its signs, the specification of its rules for forming permitted configurations, or in its higher-level structure, are in frequent practice resolved through symbolization, by going* outside *the original system.*

That is, the signs or configurations of signs in the first system must be *interpreted* by means of another representational system in order to decide the ambiguous case. *Within* the original system the situation remains (and *should* remain) ambiguous. The only alternative to acceptance of this principle, it seems to me, is to *extend* the original system considerably, constructing ever-more-complex sets of highly specialized rules to be included within the original system, as a substitute for the power of the second system.

The examples in the first section can now be looked at again, not just as representational systems, but as *representations*—that is, representational systems *symbolizing* other representational systems. For instance the symbols p, q, r, . . . in the sentential calculus can stand for (i.e., symbolize) declarative statements in a natural language; the symbol $\wedge$ can stand for "and," the symbol $\rightarrow$ for "implies," and so forth. Now the possibility of assigning truth values to natural language statements provides us with a model (a represented world, or more precisely a second representational system together with a symbolization relationship) for the system of logic.

Of course, this symbolization gives us an *imperfect* model, because the semantics of natural language is often at variance with formal logic. It is not a theorem that $(p \rightarrow q) \rightarrow (\sim p \rightarrow \sim q)$, but in ordinary usage, the statement "If you do your homework, you may go the movies," may well imply, "If you do not do your homework, you may not go to the movies." Likewise in formal logic, $p \wedge q$ is equivalent to $q \wedge p$, but in ordinary usage, "He ran into the road and he was hit by a car," is not equivalent to, "He was hit by a car, and he ran into the road." Such imperfections in the symbolization of "everyday" English by symbolic logic can account for some of the difficulties students have in learning the subject—and this points to the importance of characterizing and studying cognitive representational systems if we wish to develop useful models for learning and problem solving in mathematics.

---

[2]I would argue that for any grammatically correct English sentence, it is in principle possible to construct a context in which the sentence would be semantically admissible as well. However, this is not critical for the present discussion.

More important than the ambiguities or imperfections that may or may not be present in a symbolization, is the fact that there is no *necessity* to any particular choice of symbolization. Part of the conceptual revolution in mathematics mentioned above was the realization that the *interpretation* of mathematical axioms and theorems, as well as the interpretation of "logical languages," was wholly independent of mathematics or logic as purely *formal* systems. Thus "lines" in Riemannian geometry could just as well mean great circles on a sphere in Euclidean space. Therefore when one considers mathematical systems or logical languages as purely formal structures (i.e., as representational systems and no more), one can ask questions about their *consistency* (whether it does or does not happen that a configuration and its negation are both theorems) but not about their *truth*. Within the structure, the word "true" is inappropriate; we have instead the (possibly inequivalent) notions, "provable from the axioms using admissible procedures," "negation not provable," or "introducing no new contradictions." Axioms are no longer to be regarded as "self-evident truths" about nature, human language, or any other domain, but as arbitrary (and hopefully, nonmutually contradictory) assumptions. The historical notion of an axiom as a "self-evident truth" depended on an *interpretation* of the representational system of mathematics by means of another representational system external to that of mathematics, which the mathematics had been tacitly taken to symbolize.

Once this realization occurred, it could be observed that permissible configurations in mathematics—well-formed statements about numbers, chains of reasoning from axioms to theorems in accordance with well-defined rules of procedure—can themselves be represented by numbers. Thus mathematics can be regarded as a *self-referential* system of representation, and it was possible for Gödel to demonstrate his remarkable results (see Nagel and Newman, 1958; Hofstadter, 1979, Kline, 1980).

A relationship of symbolization between two representational systems may, as noted, be ambiguously specified. We often refer to such symbolization as *implicit* or *context-dependent,* as in the semantic interpretation of speech. To understand how such ambiguity is resolved, that is to examine in detail how the symbolization depends on contextual variables, it is helpful to try to characterize what is meant by "the context" in relation to the two representational systems under discussion. In general, I would assert the following alternatives:

> *The contextual conditions needed to resolve ambiguities of symbolization may be expressed in terms of either the first or the second representational system, in terms of the relationship of symbolization itself, or (most interestingly) in terms of still other representational systems* external *to either of the first two, for which a symbolization relationship exists with at least one of the original systems.*

For example, ambiguities in the symbolization of visual images through spoken words and sentences may be resolved by means of the following types of contextual conditions: the spoken verbal context (concomitant words and phrases), the visual image context (concomitant images or operations on them), the symbolization context (concomitantly established symbolic relationships between words and images), or the context of an external representational system (concomitant symbolization of words and/or visual images by other means, such as characters in mathematics).

Thus the rules of symbolization are not static. They can change as changes occur in representational systems *outside* of those that are symbolizing each other. They can also, of course, be held constant for purposes of discussion.

The representational systems of formal logic and mathematics discussed above illustrate a frequently observed pattern in the *historical evolution* as well as the *cognitive development* of representational systems noted during the 1984 Montreal conference, characterized by three stages:

(1) First is the act of symbolization (the semiotic function), where signs or configurations are taken to stand for aspects of a previously established representational system (cf. Piaget, 1969, p. 31).

(2) This is followed by a period of development of the structure of the new representational system, patterned after or driven by structural features of the previously established system.

(3) Finally, the new system is *separated* from the originally symbolized system. It is seen that there is no necessity to the original symbolization, that sufficiently many ambiguities in the new system can be resolved without reference to the original system, and that alternative symbolic relationships with other representational systems are possible.

Let us consider for a moment the distinction between structure and function as expressed, for example, in different schools of linguistics. If we take spoken language as one representational system, and consider it as symbolizing nonverbal, internal cognitive representational systems in individuals, we can understand the functionalist approach to linguistic theory as seeking to rely as heavily as possible on stage (2) above in explaining the grammar of natural language, and the structuralist approach as relying on stage (3) to the maximum extent (Chomsky, 1965; Diver, 1981; Reid, 1985). It appears that one can make such a distinction in almost any situation where one representational system has evolved or developed through a symbolic relationship with another. This way of expressing the differences between these two schools of thought resolves many apparent contradictions between them.

Before leaving this section, let us consider the remaining examples of the previous section as *representations*.

For the Turing machine (Example 2), we can let the set of signs that can be entered on tapes be the characters of a second representational system, and the tapes themselves be the configurations in that system. The correspondence between signs on the tapes and (contingent) transitions of the machine permits the world of tapes to symbolize the machine, or conversely the machine to symbolize the world of tapes.

A measuring device, such as the thermometer discussed above (Example 3) represents the physical state of the system whose temperature is being measured. A physical state to a physicist might be a statistical ensemble of configurations (positions and momenta) of molecules; but for everyday purposes, it might be described instead by sensations such as hot, warm, cool, or cold. The correspondence (however precise or fuzzy) between physical states (however specified) and instrument readings, constitutes the symbolization. Conversely, one might take the more unusual perspective and say that the physical state represents the thermometer.

Finally, we may note that a problem state-space (Example 4), regarded as a representational system, can be considered to symbolize certain features of any of the following representational systems external to it: (a) a system of *problems* posed verbally, iconically, by means of a concrete apparatus, or some other way, (b) a system of *algorithms, strategies,* or *heuristic processes* which can generate paths or sets of possible paths within the state-space, or (c) a system of overt *behaviors* of problem solvers. All three of these interpretations of the state-space are discussed in my earlier work on task structure variables (Goldin, 1984a).

With the above definitions and examples in mind, let us turn to the question of interpreting cognitive representational systems.

COGNITIVE REPRESENTATIONAL SYSTEMS AND PROBLEM SOLVING

In previous papers I proposed a model for competence in mathematical problem solving based on four "higher-level languages":

(1) a *verbal/syntactic* system,
(2) nonverbal systems for *imagistic* (visual-spatial, auditory, kinesthetic) processing,
(3) *formal notational* systems of representation, and,
(4) a system for heuristic planning and executive control, abbreviated as *planning* language.

It is my present view (Goldin, 1985b) that a fifth "language" is needed in order to obtain a model that can effectively simulate the structure of human competence in mathematical problem solving:

(5) an *affective* system which monitors and evaluates problem-solving progress.

The main features of this model are summarized in Figure 12-1.

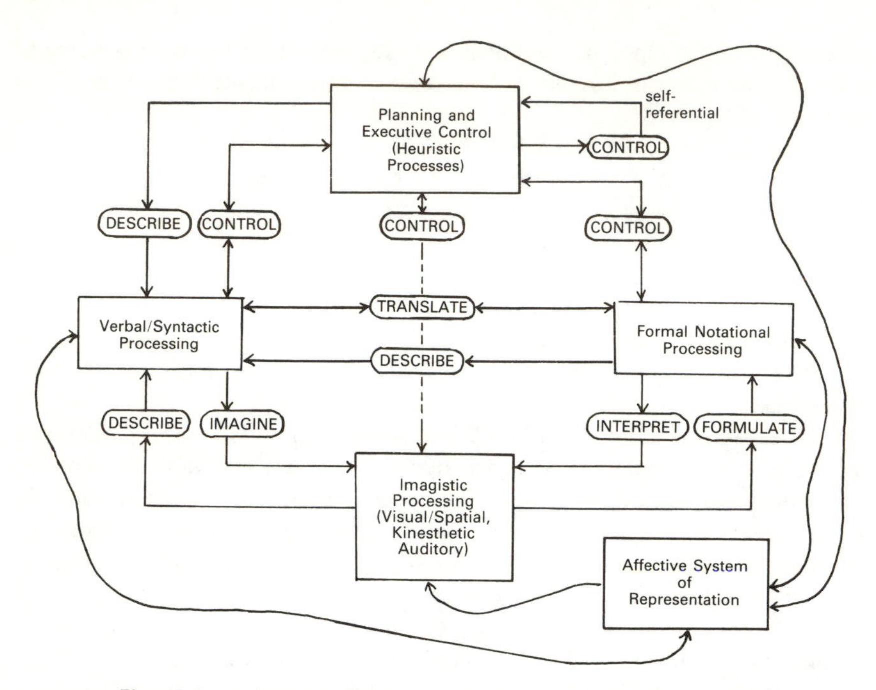

Fig. 12.1. A Model for Competency in Mathematical Problem Solving

Here I would like to interpret these higher-level languages as internal, cognitive representational systems. Let us note that we expect considerable similarity in structure between the representational systems constructed to describe the competencies of different individuals in the same culture. We also expect some similarity between individual internal representational systems, and external representational systems which are directly observable—particularly because the behavior we want to explain by means of the internal system is manifested externally. However, we must always be careful to distinguish between systems intended to model the internal states of a problem solver, and those intended to represent external behaviors or external problem states.

In this section, I would like to consider first the epistemological issue of what these languages actually talk about, symbolize, or represent, and then the ontological issue of what configurations in such representational systems really are. Next I shall look briefly at each system, and ask about the senses in which it is meaningful to call it a language.

From the previous discussion, we see that is is possible (at least in principle) to consider each language as a representational system having its own signs, configurations, and higher-level structure. There is also no difficulty in permitting these representational systems to symbolize *each other,* as appropriate. The epistemological difficulty enters when we ask whether the world of experience (i.e., any or all of the five representational systems under discussion, or some other internal representational system) does or does not symbolize a world of external reality which influences its development. Without pretending to offer a new or unique answer to the long-standing philosophical question of the existence of a real world (the "problem of transcendent inference"), let me suggest that the difficulty here is not an essential one by offering two possible solutions to it:

Solution 1: Imagistic representational systems may be taken to include configurations constructed from characters corresponding roughly to what philosophers have called sense-data. Within such systems, a variety of transitions from configuration to configuration are possible ("the world of *imagined* experience"). When we make statements about the real world, as opposed to statements about our experiences of it, we are simply abbreviating large numbers of statements about *contingent* experiences. We are not saying that the representational system called real world has any more absolute existence than representational systems we might call perceptual or imagistic. We are, however, noting that the contingent experiences associated with real-world statements (e.g., "If I hold my finger in the candle flame, then it will hurt," or, more precisely, "If [nonverbal visual-motor experiences associated with holding my finger in the candle flame], then [nonverbal experience called pain in finger]") are not subject to the same kind of *executive control* (response to planning language configurations) as imagined experience. I can choose to imagine holding my finger in the candle flame and not feeling pain, but I cannot normally choose to experience this. Likewise, real-world contingent experiences do not display the unreliability or nonreproducibility of the contingent experiences we associate with dreams or hallucinations. In this view, what I as an individual call the real world is itself a representational system—that is, a useful construct used to symbolize the world of experience, and evidencing the three developmental stages for representational systems discussed above.

Solution 2: This approach relies more heavily on the distinction between the unique epistemological perspective of the reasoning entity (I, myself), and the interpretation of representational systems that I infer others to have. Let us accept the premise that personally, I must regard my own perceptual or imagistic configurations as part of an internal representational system that symbolizes nothing more fundamental than itself; that is, for *me* there is no real world beyond the world of personal experience. But just as I have no direct access to a real world, I also have no direct access to anyone else's world of experience. Since I am trying to construct a theory of the psychology of problem solving which explains other peoples' competencies (in order to account for my personal observations of problem-solving behavior and my readings of other researchers' observations, and to improve instruction in mathematics as I observe it to take place), it is perfectly consistent for me to include configurations from my *own* imagistic representational system in the theory. Then, when I talk about the influence of the real world on some else's cognitive representational systems, as if those systems actually represented or symbolized an external reality, I am merely abbreviating observed or conjectured contingencies between experienced sequences of configurations in my own imagistic representational systems, and changes of state in (my model for) the other person's cognitive representational systems.

It seems to me that these possibilities, together with the observation that cognitive representational systems can represent each other, answer the objections of von Glasersfeld and Mason, justifying the use of the term representation and even the term real world in the present context.

Now let us touch on the ontological question of what configurations in internal, cognitive representations "really are." We shall see that this question has its parallel in the theory of the physical sciences.

A goal of cognitive psychology highly relevant to mathematics education is the construction of scientific models which can account for observed human problem-solving behavior. Most cognitive psychologists have come to reject the possibility of expressing such models exclusively in terms of directly observable stimulus situations and behavioral responses, as once was advocated (e.g., Skinner, 1953). Nevertheless, the adequacy of any scientific model must ultimately be tested through increasingly well-controlled observations (in this case, observations of behavior) to test the validity of its predictions. A number of theoretical constructs thus enter into a model for problem-solving competence based on cognitive representation.

One of these is the well-known idea of *competence,* as distinct from performance. A competency describes the capability of an individual to perform successfully on a class of tasks. Attribution of a competency does not necessarily imply that the individual actually performs successfully every time the task is encountered. The model on which I am working is an attempt to construct a description of such competencies which can help in understanding observed problem-solving behavior, but it does not seek directly to describe performance. To begin with, then, we must understand even the fairly uncomplicated idea of a competency not as something real or existing in the problem solver, but as the researcher's way of representing classes of potentially observable behavior.

Cognitive representational systems are, in turn, theoretical constructs devised by the researcher to help describe or account for competencies. When we attribute certain cognitive representations to an individual, speaking as if the individual really has such representations and as if we are modeling or depicting them, we are using a kind of epistemological shorthand—in effect, we are describing certain classes of related competencies which in turn, attributed to the individual, merely summarize behaviors which are potentially observable in that person. The value of a model based on cognitive representation lies in its promise for rendering the extraordinarily complex behavior patterns associated with mathematical learning and problem solving intelligible, predictable, and teachable.

Thus, cognitive representations are still further removed than are competencies from empirically observable phenomena and, of course, we may need to devise or use technical notations of various kinds to express them. Implicit in the question of the "actual existence" of such representations in the individual is, I think, the idea that they might somehow (at least in principle) be measured or observed more directly than through the study of behavior—for example, through an improved understanding of the human brain and nervous system. If we are willing to ascribe physical "reality" to biological states of the brain, and if we grant that detailed knowledge of such states could in principle account for behavioral phenomena, then the reality of cognitive representational configurations becomes a question of identifying such configurations with biological configurations of neurons. Perhaps this will eventually happen. However, I think it is more likely that the kinds of cognitive representational systems mentioned above will turn out to be gross simplifications, describing only certain macroscopic features of neural states; while the biological description, if it becomes possible, will be far more complex (and much less useful to educators).

There is a close analogy between the question of the reality of cognitive representations, as discussed above, and the parallel question of the reality of certain constructs in physical theories. Consider, for example, the notions of an electron, an electromagnetic field, or a quark. As purely theoretical constructs, these were introduced to account for certain observed phenomena in physics. In the case of the electron, more direct measurements later became possible which made it easier to say that it "really exists," even though one still could not see an individual electron—for one thing, its size is much smaller than the wavelength of visible light. Electric and magnetic fields were introduced into physics as convenient mathematical idealizations, to describe the theoretical forces that would be exerted on infinitesimal charges—but of course, there *are* no such infinitesimal charges. Initially regarded purely as mathematical tools, fields later came to be thought of by most physicists as "real" due to the enormous success of theories in which they enter as the fundamental physical quantities. Quarks are hypothetical particles which have never been observed in isolation and, in some theories, could never be so observed; but as entities assumed to exist only in combination with each other, they account for many of the properties of particles that *can* be observed. In each case, we see that the question of objective physical existence can be answered only by making statements about the kinds of observations that can and cannot be made in connection with the construct. Thus cognitive representations, regarded as scientific constructs to help in our understanding of problem-solving behavior, fare no worse ontologically than many other constructs of proven usefulness that enter into scientific descriptions.

I would like to conclude this section by discussing the appropriateness of the term language in referring to cognitive representational systems, an issue raised by diSessa. In particular, we may ask what it means to call the above five representational systems higher-level languages.

In using the term originally, I had in mind a meaning similar to its usage in computer science, where one distinguishes among machine language, assembler languages, and higher-level languages—that is, language as a structured symbolic system. Taking machine language to be analogous to neural events on the cellular and biochemical levels in the brain, we have already intimated that even a detailed, microscopic understanding of the workings of the human nervous system would be unlikely to yield an intelligible theory of human problem solving—any more than a machine-language version of a complex computer program written originally in a higher-level language would help us very much in understanding the structure of its capabilities. Thus, a theory of problem-solving competence, to be of practical value in education, should be expressed in languages on a level accessible, for example, to classroom teachers.

But before calling a representational system a higher-level language, I think it is fair to ask that it have at least some of the following attributes commonly associated with the term:

(a) a reasonable degree of complexity;
(b) some rules analogous to "rules of grammar;"
(c) the capability of being used to *symbolize* a reasonably wide variety of things; and
(d) use or potential use by human beings to communicate with each other.

Let us consider each of the five representational systems in turn, with respect to these characteristics.

A verbal/syntactic system of representation can be described by means of signs which are, let us say, words (written and spoken) and punctuation marks, together with: correspondences between written and spoken words, rules for tagging words by part of speech, and grammatical rules for combining words-tagged-by-part-of-speech into sentences. In addition, let us decide to include in this system some purely verbal dictionary information which may be available to the problem solver (e.g., definitions of words and phrases, synonyms, antonyms, even common usages). Our purpose is to be able to represent certain verbal processes which take place during problem solving, such as rote recall of a definition (in which the problem solver makes no use of visualization or other internal representations), *within* the purely verbal system (or between the verbal/syntactic system and the external environment); and to represent more complicated processes as information exchanged *between* the verbal/syntactic system and the other representational systems. It appears clear that all four of the above "language" criteria are met for this system.

Imagistic systems of representation are much more difficult to describe, though it seems clear that they should include visual-spatial, kinesthetic, and auditory systems in order to be able to describe problem-solving competence. For simplicity, let us consider a visual representational system. Minimally, its characters should include objects, attributes, and relations, all of which we regard as complex nonverbal cognitive constructs from sensory input. Now configurations are to be formed from these in accordance with appropriate semantical rules, which are not easy to express in a general form. For example, consider the famous sentence, "Colorless green ideas sleep furiously" (Chomsky, 1965, p. 149). While the sentence on the verbal level is grammatically correct, the imagistic configurations evoked by the sentence violate numerous semantical rules of our imagistic representational systems—"colorless" is an attribute value of "color" which *contradicts* the value "green;" "ideas" as objects do not *possess* the attribute "color" in the first place, and so forth. We are thus led to the following perspective:

> *The often tacit semantical rules which govern the images that we visualize constitute (at least in part) the* grammar *of* imagistic language.

The tacitness of these rules poses no difficulty; indeed, human competence in ordinary verbal grammatical rules is also for the most part tacit knowledge. The inconstancy and vagueness of such rules fall within our earlier discussion of ambiguity.

The visual-imagistic representational system should also include the capability of transforming one configuration into another by a variety of means—in modeling problem-solving competence in mathematics, for example, we would want to incorporate here such abilities as "mental rotation of a geometric figure." The phenomenological primitives or *p-prims* which diSessa has found to be so helpful in accounting for the cognitions of physics problem solvers, appear to me to be best described as coherent imagistic representations. One of the most interesting phenomena in the literature on mental models and misconceptions is the unusual degree to which such representations endure, even in the face of formal instruction based on models which contradict them. We thus make the general observation:

> *Coherent imagistic cognitive representational configurations tend to* persist, *despite the individual's experiences of contravening configurations in verbal/syntactic or formal notational representational systems.*

It should also be noted that *formal notational* characters can be (and are) treated imagistically by problem solvers. When we abbreviate a formal step in algebra by saying, "Bring the x over to the other side of the equation and put a minus sign in front," we are treating the formal configuration as a configuration of imagistic entities, to be manipulated visually and kinesthetically.

In my opinion, imagistic representational systems also meet all four criteria for being considered a language. They are complex, they have a grammar consisting of semantical rules (which may be more variable or context-dependent than verbal/syntactic grammar), they can represent an extremely wide variety of human experience, and certainly they are used in human communication (gestural communication, body language, figural or pictorial symbol-schemes, etc.).

Competence in formal notational systems includes the ability to use the notations conventionally described as the language of mathematics—formal arithmetic, algebra, trigonometry, calculus, and so forth, as well as "knowledge" of how to represent problem states and make moves from one state to another in non-standard problems. There would be, I think, little argument with the assertion that all four language criteria are met here.

In describing a representational system for planning and executive control, the heuristic process can be taken as a natural unit for analysis (Goldin and Germain, 1983; Schoenfeld, 1979, 1980, 1984; Goldin, 1984b, 1985b,c; Polya, 1962, 1965). Some heuristic processes have been discussed extensively in the literature. In my own work, I have based the analysis of heuristic processes on four dimensions, with respect to which the subprocesses involved in their use can be understood:

(1) Advance planning reasons for using the process;
(2) (Domain-specific) ways in which the process can be employed;
(3) The type and level of domain to which the process can be applied; and
(4) Prescriptive criteria indicating *that* the process should be employed.

Heuristic processes can not only motivate steps within the other representational systems, but they can also modify the rules of procedure within those systems; for example, a problem-solver can adopt an improved formal notation. In addition, heuristic processes have self-referential and recursive capabilities.

It is interesting that the language criterion which does not appear to be satisfied by the cognitive representational system of planning and executive control is that of "use in communication." We do communicate plans and planning decisions, to each other and to students of mathematics, but we usually do so through one of the other systems of representation—through verbal statements, or by actually taking the steps and thus exemplifying the plan. The situation reminds me of early texts in algebra, where ideas which today would be expressed easily and concisely in formal notation are instead described laboriously in words. Perhaps our inability to communicate planning and executive processes in mathematical problem solving more directly reflects the fact that our understanding of heuristic processes is still in its infancy—and we have not devised an appropriate external representational system (or *planning symbol-scheme*) to describe them very effectively.

Finally, we have the notion of an *affective* representational system playing an important *cognitive* role. In considering such as system as part of a model for describing competence in problem solving, I do not wish to focus on what might be called "global affect," such as the individual's attitude toward mathematics or self-concept as a problem solver. Instead, the intent is to be able to represent the changing states of feeling that a problem solver experiences—and expresses—while solving a problem. These states might include bewilderment, frustration, anxiety, discomfort, satisfaction, pleasure, elation, etc. They perform an important function (in highly competent, expert problem solvers, as well as in novices) in *monitoring* the solver's progress through the problem. Perhaps it is problematic whether this representational system is sufficiently complex to be considered a language, though the remaining criteria appear to be satisfied.

CONCLUDING COMMENTS

If the present model has some degree of validity in describing the structure of competence in mathematical problem solving, it suggests that the major goal of mathematics education should be to *foster the development of such cognitive representational systems*. That is, we should not merely define the objectives of mathematics instruction as teaching students to be able to solve certain classes of problems; we should instead choose problem-solving activities which contribute the most toward the development of all of the representational systems, individually and collectively.

For example, consider the mathematical concept of multiplication by -1. To express our instructional goals strictly as behavioral objectives would be to ignore the cognitive representational systems whereby we hope to achieve those objectives. We would then qualify as acceptable any internal process, however incoherent or however mechanical, leading to the correct answer. But even if we try to achieve the deepest possible understanding of multiplication by negative quantities by carefully describing all of the tacit knowledge needed to solve successfully a class of related problems, we are still permitting the tasks to determine our educational objectives—and if we succeed thereby in providing some students with a mathematics education of high quality, we have done so (in a sense) by accident. The instructional goals that I would advocate in teaching this concept, based on all of the above discussion, would be expressed in terms of achieving maximal development of the student's internal formal representational systems, imagistic systems, and so forth (behaviorally, we are thus fostering transfer of learning). One would then judge various "embodiments" of the concept (reflection through the origin, reversal of direction, debt versus profit, formal embodiment of the relationship to the distributive property, etc.) mainly by their value in achieving cognitive representational objectives, rather than by their immediate value in helping the student remember a rule.

Let me close with the following possibly controversial comment. One *negative* effect of the revolution in mathematics, whereby the possibility of mathematics as a *purely formal* endeavor was recognized, has been the widespread adoption among mathematicians of the idea that the *formal* system is what the mathematics really *is*—and that therefore other representational systems symbolized by the formal system should be suppressed as early in the pedagogy as possible. We have seen in the above discussion, however, that *no representational system is more real than any other,* and that while the relationship of symbolization is not a *necessary* one, it is often the *raison d'être* of a representational system. I think that in the distant future, the twentieth century may be seen as a formalist period in mathematics that was eventually superseded—and that, one day, we will have a body of mathematics in which ambiguity is commonplace, and in which *imagistic* configurations and *heuristic* configurations are as acceptable as formal notational systems. Then, perhaps, pure mathematicians will no longer see their main goal as the elimination of these from their development of mathematics.

REFERENCES

Chomsky, N. (1965). *Aspects of the Theory of Syntax.* Cambridge, Mass.: M.I.T. Press.

de Saussure, F. (1959). *Course in General Linguistics,* New York: Philosophical Library.

diSessa, A. (1982). Unlearning Aristotelian Physics: A Study of Knowledge-Based Learning. *Cognitive Science 6,* 37–75.

diSessa, A. (1983). Phenomenology and the Evolution of Intuition. In Gentner, D. and Stevens, A. (Eds.), *Mental Models* (pp. 15–33). Hillsdale, N.J.: Lawrence Erlbaum Associates.

Diver, W. (1981). On Defining the Discipline. *Columbia University Working Papers in Linguistics 6,* 59–116.

Goldin, G. A. (1982). Mathematical Language and Problem Solving. *Visible Language 16,* 221–238.

Goldin, G. A. (1983). Levels of Language in Mathematical Problem Solving. In Bergeron, J. C. and Herscovics, N., Eds., *Procs. of the Fifth Annual Meeting of PME-NA* (Int'l. Group for the Psych. of Math. Educ., N. American Section), *Vol 2,* (pp. 112–120). Montreal: Concordia University Department of Mathematics.

Goldin, G. A. (1984a). Structure Variables in Problem Solving. In Goldin, G. A. and McClintock, C. E., Eds., *Task Variables in Mathematical Problem Solving,* (pp. 103–169). Philadelphia: The Franklin Institute Press. Prev. pub. Columbus, Ohio: ERIC Clearinghouse for Mathematics, Science, and Environmental Education (1979).

Goldin, G. A. (1984b). The Structure of Heuristic Processes. In Moser, J. M., Ed., *Procs. of the Sixth Annual Meeting of PME-NA,* (pp. 183–188). Madison, Wisc.: University of Wisc. Center for Education Research.

Goldin, G. A. (1985a). Thinking Scientifically and Thinking Mathematically: A Discussion of the Paper of Heller and Hungate. In Silver, E. A., Ed., *Teaching and Learning Mathematical Problem Solving: Multiple Research Perspectives,* (pp. 113–122). Hillsdale, N.J.: Lawrence Erlbaum Associates.

Goldin, G. A. (1985b). The Structure of Heuristic Processes for Mathematical Problem Solving. Paper presented at the AERA Annual Meeting, Chicago, Illinois.

Goldin, G. A. (1985c). Studying Children's Use of Heuristic Processes for Mathematical Problem Solving through Structured Clinical Interviews. In Damarin, S. K. and Shelton, M., Eds., *Procs. of the Seventh Annual Meeting of PME-NA,* (pp. 94–99). Columbus, Ohio: Ohio State University.

Goldin, G. A. and Germain, Y. (1983). The Analysis of a Heuristic Process: "Think of a Simpler Problem." In Bergeron, J. C. and Herscovics, N., Eds., *Procs. of the Fifth Annual Meeting of PME-NA, Vol. 2,* (pp. 121–128). Montreal: Concordia University Department of Mathematics.

Goldin, G. A. and Landis, J. H. (1985). A Problem Solving Interview with Stan (Age 11). In

Damarin, S. K. and Shelton, M., Eds., *Procs. of the Seventh Annual Meeting of PME-NA,* (pp. 100–105). Columbus, Ohio: Ohio State University.

Greeno, J. G. (1983). Conceptual Entitities. In Gentner, D. and Stevens, A. (Eds.), *Mental Models,* (pp. 227–252). Hillsdale, N.J.: Lawrence Erlbaum Associates.

Hofstadter, D. R. (1979). *Gödel, Escher, Bach: An Eternal Golden Braid.* New York: Basic Books.

Hunt, E. G. (1975). *Artificial Intelligence.* New York: Academic Press.

Johnson-Laird, P. N. (1982). Propositional Representations, Procedural Semantics, and Mental Models. In Mehler, J., Walker, E. C. T., and Garrett, M., Eds., *Perspectives on Mental Representation,* (pp. 111–131). Hillsdale, N.J.: Lawrence Erlbaum Associates.

Kaput, J. (1983). Representation Systems and Mathematics. In Bergeron, J. C. and Herscovics, N., Eds., *Procs. of the Fifth Annual Meeting of PME-NA, Vol. 2,* (pp. 57–66). Montreal: Concordia University Department of Mathematics.

Kaput, J. (1985). Representation and Problem Solving: Methodological Issues Related to Modeling. In Silver, E.A., Ed., *Teaching and Learning Mathematical Problem Solving: Multiple Research Perspectives,* (pp. 381–398). Hillsdale, N.J.: Lawrence Erlbaum Associates.

Kaput, J. (1986). Towards a Theory of Symbol Use in Mathematics. In Janvier, C. (Ed.), *Problems of Representation in the Teaching and Learning of Mathematics,* Hillsdale, N.J.: Lawrence Erlbaum Associates (accompanying chapter, this volume).

Korzybski, A. (1958). *Science and Sanity: An Introduction to Non-Aristotelian Systems and General Semantics* (4th ed.). Lakeville, Conn.: Institute of General Semantics.

Kline, M. (1980). *Mathematics: The Loss of Certainty.* New York: Oxford University Press.

Lesh, R. (1981). Applied Mathematical Problem Solving. *Educational Studies in Mathematics 12,* 235–264.

Lesh, R., Landau, M., and Hamilton, E. (1983). Conceptual Models and Applied Problem-Solving Research. In Lesh, R. and Landau, M., *Acquisition of Mathematics Concepts and Processes,* (pp. 263–343). New York: Academic Press.

Nagel, E. and Newman, J. R. (1958). *Gödel's Proof.* New York: New York University Press.

Nilsson, N. J. (1971). *Problem-Solving Methods in Artificial Intelligence.* New York: McGraw-Hill.

Palmer, S. E. (1977). Fundamental Aspects of Cognitive Representation. In Rosch, E. and Lloyd, B. B., Eds., *Cognition and Categorization,* Hillsdale, N.J.: Lawrence Erlbaum Associates.

Piaget, J. (1969). *Science of Education and the Psychology of the Child.* New York: Viking Press.

Piaget, J. (1970). *Structuralism.* New York: Basic Books.

Polya, G. (1962, 1965). *Mathematical Discovery: On Understanding, Learning, and Teaching Problem Solving, Vols. 1* and *2.* New York: John Wiley.

Quine, W. V. (1960). *Word and Object.* Cambridge, Mass.: M.I.T. Press.

Reid, W. (1985). *Verb Number in English: A Functional Explanation* (manuscript submitted for publication).

Russell, B. (1919). *Introduction to Mathematical Philosophy.* London: George Allen & Unwin.

Russell, B. (1940). *An Inquiry into Mearning and Truth.* London: George Allen & Unwin.

Schoenfeld, A. (1979). Explicit Heuristic Training as a Variable in Problem Solving Instruction. *Journal for Research in Mathematics Education 10,* 173–187.

Schoenfeld, A. (1980). Teaching Problem Solving Skills. *Am. Math. Monthly 87,* 794–805.

Schoenfeld, A. (1984). Heuristic Behavior Variables in Instruction. In Goldin, G. A. and McClintock, C. E., *Task Variables in Mathematical Problem Solving,* (pp. 431–454). Philadelphia: The Franklin Institute Press.

Skinner, B. F. (1953). *Science and Human Behavior.* New York: The Free Press.

Skinner, B. F. (1957). *Verbal Behavior.* New York: Appleton-Century-Crofts.

Skinner, B. F. (1966). An Operant Analysis of Problem Solving. In Kleinmuntz, B. (Ed.), *Problem Solving: Research, Method, and Theory,* (pp. 225–257). New York: John Wiley.

13 Conceptions and Representations: The Circle as an Example

Claude Janvier
Université du Québec à Montréal

ABSTRACT

This chapter addresses the issue of representation as an internal construct corresponding to an external abstract configuration. It attempts to extend DiSessa's phenomenological primitives to mathematics (more precisely to the notion of circle). It was originally inspired from a research project conducted in France by Artigue and Robinet (1982).

The chapter examines various acceptations of the word representation. The notion of circle is used as an example. Primitive conceptions are presented together with two tasks aimed at probing their presence or their evolution. Two computer programs bringing forward the development of primitive conceptions are described. Negative results are analysed in view of the intrinsic difficulty of pinning down such elusive mental constructs. The conclusion stresses the importance of the research project while computers start to be used to enlarge children's universe of experimentation (microworld).

Lesh (1979) and Janvier (1980) (see Chapter 3) studied the translations involved between various representations of the concepts of rational numbers and variables. In Janvier (1983) (see Chapter 7), we have tried to extend the notion of representation so as to encompass an internal counterpart to the visible sign. We showed that the concept of function was only unique from an axiomatic perspective. In other words, it could be broken into several nonoverlapping semantic domains, such as variable, transformation, sequence, isomorphism. Differences between the domains variable and transformation were pointed out based on psychological grounds. This chapter is aimed at studying further the notion of representation considering primarily its internal component. The Montreal Conference on Representation forced us to re-examine the theoretical background of a research project that was being carried out at that time.

In diSessa (1981) (Chapter 9), we find an excellent analysis of a few phenomenological primitives (p-prims) such as springiness, squishiness, dying away. . . , which he defines as "recognizable phenomena (basic in nature)[1] in terms of which they (the students) see the word and sometimes explain it." He highlights in his conclusions two ways by which they are involved in "expert" thinking: "P-prims serve as elements of analysis, we might say models, which partially explain and provide rapid qualitative analysis for similar but more formal ideas; the recognition of a p-prims can serve as a heuristic cue to other, typically more formal analyses."

Reasoning expertise is arrived at (according to diSessa) through the construction of a priority system that links the p-prims between themselves and to the "textbook" concepts.

We believe that diSessa p-prims are very close to the "imagistic configurations" of Goldin (1983) (Chapter 6). It seems likely that expert thinking in mathematics and physics is not achieved, as he puts it: "by superior encoding of verbal problem in formal symbols or by more efficient processing fof formal notations, or by much better and more sophisticated heuristic plans."

In fact, the texts show that the three of us agree on the fact that mathematical competence can be most attributed to either the construction of a superior imagistic representational system or to a rich system of p-prims "organized by" a flexible priority system. The present chapter is aimed at (1) analyzing further the internal component of a representation, (2) describing tasks that probe two conceptions of circle, and (3) providing the preliminary results obtained.

Representation, Conception, Concept

Let us first of all clarify the terminology. As noted in Denis and Dubois (1976), the word representation has roughly three different acceptations in the psychology literature.

At first, representation means some material organization of symbols such as diagram, graph, schema, which refers to other entities or "modelizes" various mental processes. We recognize the usual domain of signifiers that refer to inaccessible "signifieds."

The second meaning is much wider. The word according to various schools of thought has several closely related acceptations that all refer to a certain *organization* of knowledge in the human mental "system" or in the long-term memory. In fact, one singles out in this meaning the more or less raw material on which cognitive activities are based. In certain cases, representation can be identified with concept. In other, they are the ingredient from which they are formed.

The third meaning refers to mental images. In fact, it is a "special case" of the second one. The distinction deserves being made because of the importance

[1]Our comment.

of the research of this domain and also because of its clear theoretical framework.

In fact, I would classify diSessa's p-prims as belonging to the second category as well as Goldin's imagistic configuration, which could also be assigned to the third meaning. As we did in Janvier (1983), we use the word *schematization* or *illustration* to refer to the first meaning and for the first time make use of a more general vocable, namely *conception,* to refer to the second meaning.

We find in the literature expressions such as preconception or preconcept, which certainly relate to the same reality. We have preferred the word conception to the word concept to stress its independence with respect to an organized theory. Indeed, several authors view the concept as the mental counterpart of entities preexisting (or existing) in a theory or within an explanatory model. In other words, concepts belong to *science* before getting a psychological status; they have a strong cultural connotation and are expected to be mental constructs that are shared by people in order to make communications possible. Conceptions as we understand them develop mostly outside organized theories as our following examples show.

As for the prefix ''pre,'' we have avoided using it because we have deliberately decided not to stress a dubious anteriority in time. In fact, several protocols (Clement, 1979 and Lochhead, 1979), for instance) show that conceptions are essentially dynamical. They are very elusive objects that are difficult to distinguish from the rules of ''*natural reasoning*'' acting on phenomelogical stimuli in conjunction with learned principles. In fact, we regard much more a conception as an *evolving* entity inside a learning sequence than as a preexisting one that determines the students' responses.

What preexists is mostly a natural logic producing rules for induction and deduction. We interrupt our theoretical considerations and present an example on which future comments will bear.

Conceptions of Circle

The notion of circle is among the first ''geometrical shapes'' children recognize. As Bialystok and Olson (1983) put it (in their own terminology):

> *round* is initially part of a perceptual recognition routine which, when combined with the significance or meaning *rolls* and *bounces* and physical object yields the concept *ball.* When the feature *round* is isolated from that object to become an object of perception in its own right with its distinctive meaning, it has become a spatial concept. The new concept *round* is equivalent in status to the concept *ball* in that it can be uniquely referenced and retrieved, manipulated, transformed, and imagined as an object of thought, and used as a basis for categorization. p. 6

As clearly stated in the last line, the aim is *categorization.* In fact, most studies in psychology that are reviewed do not involve ''advanced'' processing or handling of concepts or conceptions in relation to scriptural patterns. Roundness should, in our opinion, refer more to phenomelogical or experimental features (used as basis in "reasoning").

The textbook concept of circle that is learned at school must articulate in some manner with the initial gestaltist concept. The instruction or school concept is constructed with stencil or compass. The concept may become more analytical according to the type of instruction.

Bialystok and Olson points out rightly that one can recognize complex visual patterns and fail to identify the features that gave rise to the recognition. In other words, one may *use* spatial information to recognize objects without being conscious of the features used.

The process of perception (and we agree with them) is guided by prior frames and is more active than passive. This is a well-known fact. To explain this active role (and other reasons) they stand for a theory of *categorical representation of space* mainly achieved through the construction of "categorizing" propositional form of the type: predicate (referent, relation) such as "on (ball, carpet)." Let us note that in fact the usual teaching of geometry that mostly stresses the main features of objects with words plays this role of guidance in the perception.

We personally consider that *consciousness is too often "equated"* with verbal explanation and word (linguistic) expression. In fact, most of our prior research in the *reading of graphs* has precisely shown the great importance of word as cue and "beacon." However, consciousness can also relate to sequences of actions or processes with a minimal use of words. In other words, a first search for explanations might only elicit the articulation of a few basic actions or images via a minimal of tying words. This is exactly where we find "room" for the notion of conception. Indeed, for more advanced (than recognition) processing, judgments are based on present and prior experience that are used in conjunction with some "natural logic" as our following examples show.

Let us note before that we are very far from the notion of *representation* as a symbolic object (first meaning) and that we deal with conceptions (representations, second meaning) associated with *abstract* written sign (and not totally symbolical). In fact, one could try to extrapolate using analogy before experiments are achieved in the field of conceptions associated with signs more symbolic in nature.

A Basic Research

The experiment described next stems from an interesting research project reported in Artigue and Robinet (1982). They derived a set of conceptions from interviews based on several items and from work done in classrooms, that they relate more or less in a one-to-one fashion to 11 correct definitions. However, their report does not tackle the question of characterizing the notion of conception. At the most, they point out that the richness and complexity of "viewpoints'' can be associated with objects otherwise uniquely defined in textbooks. They note also that ''viewpoints'' are often fragmentary as we see later. They also distinguish three criteria used in comparing the definitions and consequently (for them) the conceptions: (1) the pointwise–global opposition, (2) the static–dynamic spectrum, and (3) the explicit reference in the definition to basic features such as the center, the diameter, and the radius as length and as a segment. As for the *concept of circle,* the definition that is used in textbooks evolved with the appearance of set theory. Traditionally, circle was defined as a curve such that all *lines* drawn to the curve from a common point located inside the curve would be of equal length. Recently, the curve became a set of points and the equal distance condition was simplified into ''all at equal distance from an inside point called the center'' (note that the radius as segment ''vanishes''). The concept certainly stresses the center as a major distinghising feature.

Two Tasks To Probe Conceptions of Circle

The Three-Ring Puzzle. The pupil is given an envelope containing all the parts but one of three cardboard rings each cut into seven parts (see Fig. 13.1). The pupil is told that a part is missing but does not know from which ring. Actually, the missing part belongs to the big ring. The aim of the game is to make the puzzle. The further analysis of protocols deals primarily with the method used to complete the big ring and with the general strategy utilized in making the puzzle. There are materials on the table: some blank paper, some drawing paper, a graduated ruler, a compass, a pencil, and scissors.

The Spiral Puzzle. The envelope contains the parts of a cardboard spiral (about 2 cm wide). The pupil is asked to make the puzzle and the method used is mostly observed and analyzed.

Those two tasks were used in preliminary interviews in order to verify the presence or the evolution of circle conceptions in the strategy used to make the puzzle. Questions were only of the kind: Why do you reject this part? Why do

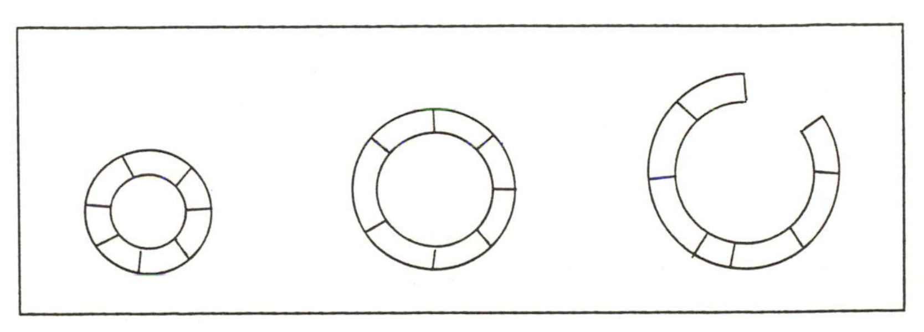

Fig. 13.1 The three-ring puzzle

you keep this one? Is this really a circle? Why? Those interviews were conducted by graduate students of mine.

Prior experiments had also been conducted in France with *recognition tasks* involving figures resembling circles but that were not. Pupils had to explain why certain figures were not circles. In that case, Robinet and Artigue mainly noted the use of school concepts in order to partially characterize circles. The idea of *symmetry* was recurrent. We may regard this rearrangement of school concepts as a conception, but we must note in that case that a conception consists in a partial characterization expressed as a necessary (but not sufficient) condition for a closed curve to be a circle. The symmetry in all directions does not have to be expressed.

The idea of *unequal diameter* was also frequently expressed. By the way, diameter was more regarded as splitting the circle into two parts than as a double radius.

But with the puzzles, we observed that school concepts are not frequently used. Actually, expressions such as "turning even," "turning the same" were frequently heard. More precisely, the French expression used was "rentrer," which at the same time involves the idea of "being bent" and "turning." In fact, only a motion with the head or the fingers can reveal whether the arcs considered are apprehended *globally* (and are looked at as differently bent), or they are more dynamically examined and the curves are found more or less sharp.

Conceptions and the Tasks

The preliminary interviews allowed us to define more clearly the conceptions in relation to the tasks and to plan an experimental design. We decided to make the following hypotheses.

The completion of the incomplete ring might be achieved using a compass in which case the textbook concept prevails in the reasoning. If another long piece is used and correctly placed to cover the gap, we can suppose then either a global curvature conception is used or the *homogeneity under rotation* is the conception at work.

This conception stems from the remarkable fact that circles with straight lines are homogeneous in all their parts in the sense that they can slide on themselves without any part showing off. We then speak of (1) a global curvature conception, (2) a dynamical curvature conception, and (3) a homogeneity under rotation conception.

The methods used to assemble the rings are expected to reveal the *global curvature conception* if the strategy used consists in superimposing the arcs to check if they "fit." On the other hand, if they are connected end to end so that they appear to turn the same way, we can suppose (and only suppose) that a more dynamic curvature conception is being used or developed.

As for the spiral, we consider that the "solution" would mostly reveal the use of the dynamical curvature conception.

Computer Programs Developing Conceptions

Because conceptions are mostly based on prior experience and mental images, we thought about trying to induce or develop circle conceptions. Both programs finally look like games and the learning can easily be seen as simply incidental. This is the way we think conceptions should develop, more implicitly than explicitly.

Moving Round the Arc into a Circle. The development of this program is an interesting story by itself but we limit ourselves to present the end-product. The central idea of this program is the rotation of an arc (of circle) achieved by turning the knob of a paddle clockwise and counterclockwise. The relation with the *homogeneity under rotation* conception is patent. However, to insert this "turning" action into a meaningful task we impose conditions on the circle obtained with the paddle.
It is required to be tangent to two lines appearing on the screen (see Fig. 13.2). The arc can be moved sideways, upwards, and downwards before it is completed. As soon as the knob is activated the resultant circle is fixed. Various levels of difficulty are introduced and a score is given by placing the arc differently at the beginning or by replacing the two lines, with one line and a point (version 2) or two points (version 3).

Riding with Your Radius of Curvature. This second computer program is not completed yet but it is aimed at developing a dynamic curvature conception of a circle. With a few constraints making it into a game, a car will be driven with a joystick and at any instant the radius of curvature will stick out perpendicular to the motion. If the car is left alone at any moment it goes on along a circular path.

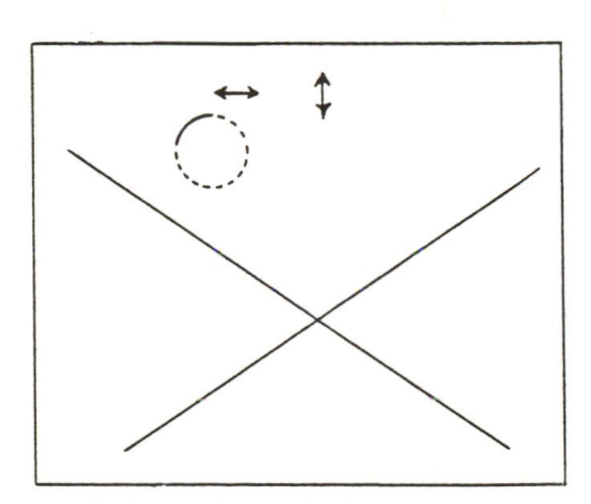

Fig. 13.2
Moving round the arc into a circle tangent to the lines.

Description of the Experiment

We have so far carried out a small-scale experiment involving the three-ring puzzle and the moving-round-the-arc program. We spent a day in a school and worked with 12 groups of two pupils aged 13–14 for one half and 15–16 for the other half. Unfortunately, we lost (for technical difficulties) two videotaped interviews and so are left with 10.

The interviewing technique is strongly inspired from Balacheff's (1981) "pupils' interaction technique." We consider it pointless to try to check how effective the conception is prior to the treatment because it might contribute partially to develop it in an unexpected manner. We perfer to put forward the hypothesis that the conception is roughly inactive at the beginning and not use any pretest.

The methodology is as follows. The control groups (of two pupils) work on various problems having received the assurance that they would be next on the computer and the experimental groups work two-by-two on the computers. After 20 minutes of computer work or of problem solving, *one* pupil of the group is presented with the task. He (she) is told that his performance does not count. In fact, when he (she) will have completed the puzzle, he (she) will explain to his (her) partner what to do. His (her) partner's time will be the one that will count. When the puzzle is solved, the second one comes in and listens carefully to his friend. A chronometer is then used to time the second pupils's performance (but for us, it has no value!).

Preliminary Results

The variety of responses obtained turned out to involve several factors that the preexperiment had not allowed us to pin down.

1. The puzzle can be more or less envisaged by the pupil as a school activity that he (or she) must succeed at and not as a game. At any rate, the desire to solve the problem quickly brings about a nervous tension that is incompatible with an open search for alternative methods.
2. There exists an ambiguity in what the final configuration of the ring should look like. In fact, we naively assumed that rings should be *circular*. They happened to be formed for pupils like ellipses.
3. The explanations from one pupil to his (her) partner turned out to be no more than a restatement of the nature of the game. Pupils could not explain "in words" what they have done. A few expressions however were made that we used in order to characterize conception more fully. Remark: Next time, the interviewer will explain the game and the pupil will start his (her) method from this point.
4. The systematicity of the approach induces various degrees of complexity to the task. If only one ring is formed at the same time, we see that we have six

''different'' games. If the rings are constructed in turn, then the multiplicity of paths to the final pattern is obvious, more particularly, if the big one is tackled first and the missing part is hopelessly searched in the pile.

5. Consequently, one response is in fact a succession of quick decisions for using a strategy or switching to another, each decision being taken all the time from a different set of ''initial conditions.'' This simple fact makes the ultimate or final strategy used basically incomparable and even casts some doubt on comparisons we would be tempted to make between the approaches used to complete the big ring.

In view of those facts, results are more than correctly qualified as preliminary. We present in turn the completion of the big ring and the first strategies. As explanations of one pupil to the other appeared to have not too much effect, their responses were not analyzed.

The Completion of the Big Ring

Out of the five pupils experimental group, *three started using a compass,* two changed their minds to use a part to fill the gap, and one simply drew the missing part. The fifth one used a part of the big circle to complete (free hand) the medium side one.

Of the control group, four pupils used a compass and one drew a missing part free band on a piece of paper and cut it with a pair of scissors.

Discussion and Interpretation

We are enclined not to attach too much importance to this difference in favor of the experimental group mainly because of the great diversity observed in the pupils getting their rings done as mentioned before. In fact, in a few cases we had to intervene.

Consequently, *we believe* that the homogeneity-under-rotation conception was *not enough present to dislodge the school concept* based on compass. In a further investigation we think that pupils should play the game for more than 20 minutes and on a few occasions.

One difference appeared more clearly. One out of four grade-8 pupils used a compass as the final method to complete the circle whereas for grade 10 the ratio is 5:6.

The First Strategy

We now give a general description of the strategies observed and discuss the difficulty we have to assign conceptions to them. Eight of the 10 pupils resort at first to the puzzle-making strategy that consists in connecting the parts end-to-end. Rejection or acceptation of each part was then based (if we rely upon reactions to our interventions, preexperimentations, and comments from one pupil to the next) on either a dynamic strategy witnessed by statements such as "turn too much," "turn the same," or on a more static one even though "end-to-end" suggests motion. In the last case, the curve seems to be examined more globally. The pupils refer then to curves being "more or less pronounced" or to "not round enough,"[2] "too much round."[2] The word bend is not as such often associated with curve. We imagine that this word in English would reveal a more static approach to the curvature.

Side by side examination clearly indicates a more global approach. It was used by a few pupils after failure with an end-to-end strategy. However, it does not guarantee success because parts are not of the same length and comparisons involved difficulties when made with small parts.

Superimposition was used by only two pupils. It is not easy to pin down the conception behind this approach. Indeed, the curvature can be appreciated dynamically before the comparison is achieved or it can be perceived more globally. Moreover, it is even impossible that a "homogeneity-under-rotation" conception be used.

Only one thing remains certain: There does not seem to be connection between end-to-end and the use of compass to complete the big ring. Indeed, we observe superimposition followed by the use of compass twice and end-to-end leading three times to completion using a part.

Discussion

As we are planning to carry out the experiment once again shortly, it seems appropriate to indicate the major expected changes. As already stated, the pupils will have the opportunity to play the game several times and over a longer period of time. We will keep our original pupils' interaction methodology and, as mentioned, the interviewer will explain the rule to the second pupil. We intend not to analyze systematically the second pupil's response. However, we have already established a list of blocking points for which we hope to arrive at a systematic reaction from the interviewer. Our intention is to focus our attention in our analysis on a few similar patterns of response in order to formulate hypotheses on the coherence of conception systems and their evolution. We hope that a large number of responses will bring us those few similar ones. More attention will be paid to the actions (gestures) made by the pupil. We will wonder if this attempt to "deverbalize" our study of conception is not presumptuous because, as we said, conceptions are very elusive entities that evolved and,

---

[2]Very approximate translation of "pas assez rond" and "trop rond" that are not even correct in French.

consequently, cannot easily be regarded as present at one time. As for the computer programs themselves, we do not expect any changes.

Conclusions

Artigue and Robinet (1982) paved the way to the analysis of various conceptions related to the notion of circle. We tried to examine the status of conceptions in relation to DiSessa's and Goldin's work. Not only do we believe that conceptions play a central role in translation processes and mathematical thinking but, like diSessa, we believe that they can be considered as the basis to the development of more formal ideas sometimes appearing late in the curriculum. For example, to become a textbook concept, the notion of curvature requires being wrapped in derivatives and disguised in a vectorial language. Clearly, in a problem-solving situation the characterizations of the conceptualized notion are called into play. However, we think that the conceptions in terms of mental images or anticipatory actions also intervene efficiently.

The main questions regarding the development of conception concern the timing. In fact, our experience as a teacher has shown several times that the introduction of a simplifying symbolism frequently resulted in a difficult go-back to the original spelled-out word description. Think about logarithms, exponentiation, trigonometric functions. As soon as formulae are provided, the meaning appears to vanish. We can easily imagine that such is the case with conceptions fading out because of "powerful" concepts. That is why we advocate that conceptions should be developed at the *preconcept* stage, in other words prior to the learning of the formalized concept.

This position has strong educational repercussions. On the one hand, we agree with diSessa that they can be used as "heuristic cues to other, typically more formal analyses." But on the other hand, they could be used as the basis for a curriculum-for-all. The debate on the nature of a core curriculum is now intense in most developed countries. The right of the population to get the best must be balanced with the need for a basic instruction for all. The approach to this question is traditional in the sense that concepts, usual algorithms, and processes are the objects rearranged in diverse pattern by curriculum designers. We propose that more attention be paid to conceptions, for they would ensure the basic knowledge for all that would later be systematized and formalized.

As microcomputers invade schools and homes, we enter into a new educational era. To use a trendy word, microworlds can now be open to pupils that enable them to explore rational systems through hypothesis testing based on inductions or deductions. Using simulation and computer's pattern recognition capability, microworlds provide the pupils with an interactive support to develop "explicative" and predictive models characterized by mental images and actions (as well as reasoning!). The computer programs we have described partially fill this goal.

We hope we have illustrated how conceptions can enter into the design of software even though we have not demonstrated that they were effectively developed.

REFERENCES

Artigue, M., Robinet, J. (1982). Conceptions du cercle chez les enfants de l'école élémentaire. *Recherches en Didactique des Mathématiques, 3,* no 1.

Balacheff, N. (1981, July). Une approche expérimentale pour l'étude des processus de résolution de problèmes. *Actes du cinquième colloque du Groupe International Psychology of Mathematics Education (PME)* (pp. 278–283). Grenoble.

Bialystok, E., & Olson, D. R. (1983). Spatial categories: *The perception and conceptualization of spatial relations.* In S. Harnad (ed.), *Categorical perception.* Cambridge, Mass.: Cambridge University Press, in press.

Clement, J. (1979). Mapping a student's causal conceptions from a problem-solving protocol. In J. Lochhead & J. Clement (Eds), *Cognitive process instruction* (pp. 133–146). Philadelphia: The Franklin Institute Press.

Denis, M., Dubois, D. (1976). La représentation cognitive: Quelques modèles récents. *Année Psychologique, 76,* 541–562.

diSessa, A. A. (1981). Phenomenology and the evolution of intuition. *Division for study and research in education,* Massachusetts Institute of Technology.

Goldin, G. A. (1983, September). Levels of language in mathematical problem solving. *Proceedings of the Fifth Annual Meeting of PME-NA* (pp. 112–120). Montréal.

Janvier, C. (1980). Translation processes in mathematics education. *Proceedings of the VIth PME Conference* (pp. 237–242). Ed. R. Kaplus, University of California. Berkeley.

Janvier, C. (1983). Representation and understanding: The notion of function as an example. *Proceeding of the Seventh International Conference of P.M.E.,* Israel pp. 266–270.

Lesh, R. (1979). Some trends in research and the acquisition of the use of space and geometry concepts. *Critical reviews in mathematics education.* Material and Studien, Band 9, Institut fur Didactik der Mathematik der Universitat Bielefeld.

Lochhead, J. (1979). On learning to balance perceptions by conceptions: A dialogue between two science students. In J. Lochhead & J. Clement (Eds), *Cognitive process instruction* (pp. 147–178). Philadelphia: The Franklin Institute Press.

14 Towards a Theory of Symbol Use in Mathematics

James J. Kaput
Dept. of Mathematics
Southeastern Massachusetts University

INTRODUCTION

This chapter begins to outline a theory of mathematical symbol systems promised in Chapter 2. In the long term such a theory should (1) help frame a systematic discussion of mathematics symbol use by individuals, including correct, partial, inappropriate, and primitive use, (2) clarify the interactions between natural language and imagistic representations on one hand and synthetic symbol systems on the other, (3) enhance our understanding of the roles of symbolic representations in mathematics growth and change, and (4) guide the development of new symbol uses made possible by new information technologies. (The reader recognizes this chapter as a companion to John Mason's decidedly more poetic contribution "What Do Symbols Represent?" in Chapter 8. Virtually every issue raised there through a range of concrete examples is addressed in this chapter. As points of parallel occur, we point them out.)

Such an ambitious set of goals could not be achieved in a single book, let alone a chapter. Hence we confine ourselves to building just enough theory to support informed agenda setting.

In our earlier chapter we showed that mathematics itself is inherently representational in its intentions and its methods. In fact, after a survey of a wide range of mathematical systems all across mathematics, their relationships, and their major results, we concluded that the idea of representation is continuous with mathematics itself. However, that entire discussion deliberately postponed reference to the role of the actual symbols used to do mathematics.

But all would agree that no significant mathematical activity is possible without material forms for its expression—forms with two closely related functions:

```
  357     (8 marks for problem)
x 296
-----
 2142
3213      (18 marks for solution)
714
------
105672
```

FIG. 14.1.

the support of internal cognitive processing and communication between persons. Further, the kinds and levels of mathematics possible are greatly controlled by the types of symbol systems used to present that mathematics.

Just as natural language and pictorial systems render manageable the flow of experience, by breaking it into chunks, by denoting those chunks with symbols, and by enabling the mind to manipulate those notations selectively stripped of the nuance and complexity of their referents, so do the symbol systems of mathematics function. To be sure, they organize but certain aspects of experience—the abstract and the structural—and they do this in ways somewhat different from the "natural" systems, with a premium on succinctness, generality, and precision of reference, whereas the latter systems depend on a compensating mix of concreteness, ambiguity, and redundancy.

The potent systems of shared mathematical symbols are the product of centuries of selection and evolution across the history of mathematics (Cajori, 1929a,b). Compare the enormous power and efficiency of the mundane mathematical representations given in Fig. 14.1–14.3 to their ancient counterparts—so much information has been distilled into so few marks. Given their ubiquity in contemporary life, it is easy to overlook their importance as tools of thought and communication. And given that these forms of representation are so highly evolved, it is likewise easy to underestimate the amount of learning required to use them intelligently.

In the first section of this chapter, after introducing a few of the central ideas, mainly through examples, we are required to deal with several important philosophical preliminaries. Our attempt to develop a theory runs headlong into an unavoidable philosophical thicket. We emerge from the thicket to examine further examples of symbol use phenomena and hence to Section B, where we introduce the problem of the relation between mathematical symbol systems and the "natural" systems of natural language and pictures. Again, our tack is to center the discussion on well-studied concrete phenomena.

$$Y = 2X + 3 \quad \text{(7 marks)}$$

FIG. 14.2.

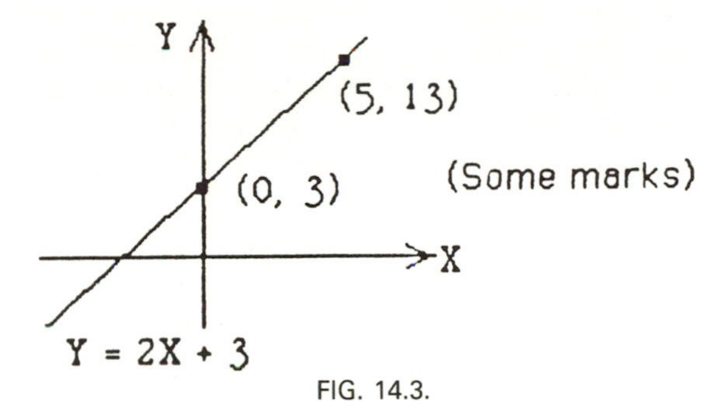

FIG. 14.3.

SECTION A: PRELIMINARIES

Outline of the Main Constructs

Most of the ideas described next can be given at various levels of formality. We attempt to strike a balance between extremes, holding the formality to that level that seems appropriate for clarity's sake, sharpening distinctions in the sequel as necessary. Some of the characterizations following have been stimulated by the work bf Goodman (1976), Gardner, Howard, and Perkins (1974), and Salomon (1979), although their primary interest was the world of art and associated media and symbol systems.

We begin with the notion of *mathematical structure* as a primitive. Examples include the various number systems, the vector spaces built on Cartesian products of real numbers, sets of functions between number systems, geometries, abstract mathematical structures such as groups, rings, topological spaces, as well as categories of such including the structure-preserving maps between them. In essence we want ''mathematical structure'' to denote any coherent chunk of mathematics that is typically found as a proper topic of study or application. Typically mathematical structures participate as subsystems or in some other role relating to other mathematical structures and seldom function in isolation.

In Kaput (1985a) and following Palmer (1977) we defined a *representation system* as involving two worlds, the represented and the representing worlds, the specification of what in the represented world is being represented, what in the representing world is doing the representing, and a correspondence between them that fixes the representation connection (see Chapter 2 for more detail). That definition was intended to help understand the many uses of the trans-theoretic term *representation,* including cognitive and perceptual representation, explanatory representation, as well as symbolic representation. Whereas in Kaput (1985a) we concentrated on the first uses listed, we now turn our attention to the last listed, symbolic representation.

A *symbol scheme* is a concretely realizable collection of characters together with more or less explicit rules for identifying and combining them. Characters can be further defined as equivalence classes of "inscriptions" (to take into account, for example, that there are many ways to write the character "a" or to draw a coordinate system), but we do not pursue the formalization further at this point. Examples include the symbol scheme of Hindu–Arabic numerals and their concatenations, and the familiar symbol scheme of two-variable algebraic expressions in, say, x and y, together with rules of combining them—their syntax. Not all symbol schemes are alphanumeric. Coordinate axes, pictures, and diagrams have other types of characters and syntactical rules.

Throughout what follows we tend to avoid explicit characterization of the rules, the syntax, of specific schemes. Readers so inclined can assume that a symbol scheme should in principle be "machine compilable." See, for example, (Kirschner, 1985) for an example of an explicit syntax for an algebraic symbol scheme. A symbol scheme, in and of itself, has no meaning, which requires a field of reference.

A *symbol system* is a symbol scheme S together with a field of reference F, and a systematic rule of correspondence c between them, perhaps, but not necessarily, bidirectional. A symbol system S will sometimes be denoted by an ordered triple $S = (S, F, c)$. We confine our attention to *mathematical* symbol systems, that is, systems where the field of reference is associated with a mathematical structure. An example is the base-ten placeholder representation system for counting numbers using the Hindu–Arabic symbol scheme (this includes the correspondence between the symbols and the elements of the system, i.e., the numbers themselves). Another symbol system with the same field of reference uses the number line scheme and the usual correspondence between the numbers and points.

Thus a mathematical symbol system is a special kind of representation system, whose represented world is a mathematical structure, whose representing world is a symbol scheme, and with a specified correspondence. In most cases, it turns out THE REPRESENTED MATHEMATICAL STRUCTURE WILL ITSELF BE A SYMBOL SYSTEM THAT CAN BE TAKEN TO BE REPRESENTING YET ANOTHER SYMBOL SYSTEM. We argue that the ontogenesis of mathematical structures within individuals, as well as historically, proceeds via the repeated consolidation and reification of actions on previously established symbol systems (where such "actions" include reflections upon the structural invariances engendered by or made salient by other of those actions). Thus, selected aspects of one symbol system become the field of reference for new symbol systems. Traces of those earlier actions and primary structures are ever present in the more advanced systems. They are the source both of mathematics' richness as well as its learnability. Their selected suppression in the next-level systems is the source of the power of mathematical symbol systems: With the earlier actions and structures crystalized into stable, concretely manipulable symbols, the mind is freed to act on or reflect upon the earlier actions and structures in new ways—perhaps leading to another cycle of mathematics building.

Informal Illustrations: Symbolization and "Entification"

We now offer a few oversimplified samples of mathematics-development phenomena that we hope a good theory of symbol systems would account for. The intent is not analysis, but setting the scope for analysis.

In the following paragraph and throughout the chapter, we make reference to "phenomenal objects" and use similar phrases. The reader sees strong parallels in meaning to Mason's "images" and "inner experiences" and to DiSessa's "phenomenological primitives," especially when we are referring to "primary referents." Each author, using somewhat different language, is attempting to account systematically for that interior aspect of thought too often neglected in educational research literature, and almost entirely ignored in the curriculum.

The characters of "one, "two," "three" get borrowed from natural language in preschool children to become the acoustic-markers for primitive recitations of counting poems that, together with correspondences with physical markers, lead to counting actions (Steffe, von Glasersfeld, Richards, & Cobb, 1983; Fuson & Hall, 1982). Such counting actions get consolidated into the adjectival numbers, means of describing certain aspects of experience (Nesher & Schwartz, 1982). These, with the stability and phenomenal "objective" nature afforded by symbolic representation, come to be discussable as having properties of their own, such as order and combinability. Hence new phenomenal objects stabilize in the child's world, whose referents are the adjectival applications—the counting actions—of the earlier system. (These new "noun" numbers happen to be denoted by the same characters as the adjectival numbers but are surely distinct from them. Adjectives do not have properties except to linguists. This dual use of the same characters is a source of much confusion, for students, teachers, and researchers.) As we see later, the adjectival property of numbers survives to control students' mathematical performance many years later.

On another root stem grow the actions of "taking parts of," especially halves, quarters, and thirds. Again, these actions, initially embodied in natural language and concrete actions, get symbolic representations of their own that in turn lead to their "entification" into phenomenological objects in their own right, capable of carrying properties and supporting new actions. Such a new action is addition, which likewise had previously become objectified from two kinds of actions: one involving counting up (using increasingly efficient strategies) based on the symbolic markers stabilized earlier (Carpenter & Moser, 1983); the other involving a combining of adjectival numbers based on the concrete entities that these adjectives describe. These "addings" (addition acts) come to be represented as an operation on the whole numbers (now having reached noun status) using the "+" character and thus becomes available—via an extension of the referent of the "+" character—to apply to fractional numbers. Note again that once "+" gets to be an object in its own right, not only can it be extended to new classes of "numbers," it can carry properties, e.g., commutativity, and support further actions, e.g., multiplication as repeated addition.

Our earlier remark about the new system retaining traces of its progenitors is well illustrated by recent work on student mental models of multiplication. Fischbein, Deri, Nello, and Marino (1985) show that the repeated addition basis of multiplication considerably and conspicuously constrains student ability to apply multiplication in practical problems when the tacit assumptions of the repeated addition model are violated, i.e., when the multiplier is not a whole number. Working in the other direction, Kieren, Nelson, and Smith (1985) have revealed how the multiplication of fractions is rooted in the composition of "taking parts of" acts (e.g., "one half of one third of").

The process of "entification–symbolization" occurs all across mathematics at all levels. Thus functions, initially understood as transformers of numbers, become objects in their own right, objects that can be added, composed, inverted, and differentiated. Then the acts of differentiation can in turn be objectified as differential operators. These likewise can be combined and otherwise manipulated in the theory of differential equations. Note that the *transformational* aspect of functions is but one aspect. It is perhaps best captured in the "$f(x) = \ldots$" notation. Another aspect is the *relational,* which is perhaps best captured by the "$y = \ldots$" notation. Each aspect has its most resonant symbolic form, and each leads in different mathematical directions. We offer more examples later in the chapter. In addition, the reader will recall John Mason's first example and follow-up discussion in Chapter 8.

An important feature of (mathematical) symbol system correspondences is their "structure-preserving" nature: The syntax of the symbol scheme must relate systematically to an identified structure of the reference field. However, it would be a mistake to formalize this requirement in a unitary way for all types of symbol systems. They vary far too much for a single characterization. Note also that the obvious and standard formalizations of structure-preserving maps between mathematical entities, e.g., morphisms in various categories or relation-preserving maps between logical systems, are too restrictive to accommodate both the wide variation across symbol system types and the psychological realities that a theory of mathematical symbol use should capture. After all, the actual use of a symbol system in practice always takes place relative to a particular person and a particular mathematical context or task.

Media and Symbol Systems

A *medium* is a physical carrier in which a symbol scheme, hence symbol system, can be concretely instantiated. Examples include pencil/paper types of static flat two-dimensional media using primarily spatial dimensions and color (a minimum of two, of course, is necessary to distinguish inscriptions from background), as well as dynamic computer and video media that utilize time as a carrier of information in addition to spatial dimensions. We deliberately include the human cognitive apparatus as a medium (suppressing attention to neurophysiology in the same way that we supress attention to hardware specifics when discussing the computer medium). A given symbol scheme, hence symbol system, may be instantiated in more than one medium—for example, most alphanumeric symbol systems can be "read aloud" by humans or machines and thus be instantiated in the (dynamic) sound medium. Moreover, media may be complex in the sense of using several different information-carrying dimensions in combination or in sequence, as with a computer using static graphics or text displays, sequences of such to simulate motion, and even sound. A given symbol scheme can be instantiated in a medium provided the medium has sufficient distinct carrying dimensions to accommodate the different features that are used to differentiate the characters of that scheme. Finally, a given medium may have important attributes independent of symbol system considerations, such as the ability to support interactive involvement or the ability to allow erasures or other transformations of existing characters, e.g., scrolling.

Note that "medium" usually is discussed as a carrier of "information," (formally definable in a Shannon sense) but that we are using it as a carrier of symbol schemes and hence symbol systems. To the extent that these provide structural "vessels" for the information, we are saying something a bit different. Of course it is also the case that the vessels themselves constitute information, so we are not contradicting that more general point of view. We are merely focusing attention on those aspects of the information that are used to organize further information.

We do not in this chapter deal with the characteristics of the different media and their relationships with the common symbol systems of mathematics. Our main reason for bringing in media at this point is to distinguish them from symbol systems, with which they are often (incorrectly) identified. The distinction between a symbol *scheme* and a symbol *system* is introduced mainly as an analytical device to help with the discussion as we progress. In practice, the syntax of a given symbol scheme is coordinated with its field of reference in order that the correspondence between them preserves certain attributes. Nonetheless, a given set of characters, and even certain rules for combining them, may participate in several different symbol systems and hence have different "meanings," i.e., referents. This situation is responsible for a whole class of symbol-use errors in mathematics, e.g., those having to do with the "=" character, which participates in several different, but related mathematical symbol systems as well as natural language as noted even in the last century by Peano. The scheme/symbol system distinction also facilitates discussion of the changes in referent for a given set of characters as a student progresses through the mathematics curriculum, e.g., the referent for "+" as one moves from arithmetic to algebra to calculus.

A Starting-Point Example: Graphing Algebraic Equations

We now close in at a finer level of detail to look more formally at an example of symbol use internal to the body of mathematics, one that emphasizes "horizontal translation" rather than "vertical development," as did the previous discussion. The example is chosen specifically to provide an uncomplicated start—uncomplicated by the linguistic issues associated with "real-world" referents for the mathematical symbols, uncomplicated by the philosophical issues associated with claims about the existence of mathematical objects (e.g., to what does the numeral "3" refer), and uncomplicated by other philosophical matters associated with whether "mental representations" represent an external world. Incidentally, there is no intention here to model an ideal curriculum or to promote a particular pedagogy, but rather to look closely at some common school symbol system acts. The reader is also encouraged to recall Mason's concrete examples in Chapter 8.

Suppose a student gets, from some source, an algebraic equation in two variables, say a linear equation in X and Y, $Y = 2X + 3$. The task is to plot this equation on a two-dimensional coordinate graph using a standard table of data. We first sort out the symbol systems used in relating the equation to a table of data.

Let us denote the symbol scheme of two-variable algebraic equations by E-2. Thus the given equation is an *application* of E-2. The central mathematical structure in this example is the set of relations on some number system N, say the real numbers in this case. We do not wish to lose sight of the fact that the numbers used in this sort of exercise are in fact being represented within a particular symbol system, in this case the base ten placeholder system. The set of numbers in this system are thus designated N-ten, and the set of relations in N (mathematically the set of all subsets of $N \times N$) represented in base ten are designated by $N \times N$-ten.

There are several other symbol systems involved in this situation as sketched in Figure 14.4: The tabular symbol system *T-2* = (*T*-2, $N \times$ *N-ten, ct*) whose symbol scheme is a table of data and whose field of reference is N × N-ten, the correspondence ct is the usual correspondence between a pair of numbers and the double entries in the table. The table, hence correspondence, involves necessarily but a finite sample of the particular subset of $N \times N$. *The symbol system (E*-2, *T-2, cs*) whose correspondence cs between the equation in E-2 and the table is based on substituting a number for X and entering double that number plus three into the corresponding place on the table. (The "*s*" is for "substitute.") It is important to recognize here that this last symbol system has another symbol system, *T-2* = (*T*-2, $N \times$ *N-ten, ct*) as its field of reference. Thus the field of reference $N \times N$-ten for the symbol system *T-2* is acting as a second order, indirect field of reference for the symbol system (E-2, *T-2, cs*) via the composite

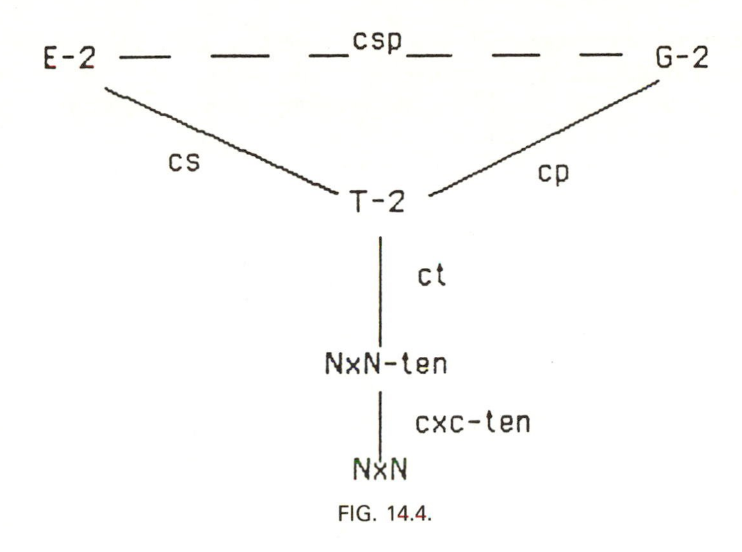

FIG. 14.4.

correspondence *cs* followed by *ct*. The difference between this "nestedness" of symbol systems and the examples of the previous subsection is the fact that the representing symbol system (*E*-2, *T-2*, *cs*) is not built up out of its referents in the style of the earlier examples. Here there is a kind of "horizontal parity" of the systems involved. The symbol system *G-2* = (*G*-2, *T-2*, *cp*) whose scheme is the usual two-dimensional coordinate plane, whose field of reference is *T-2*, and whose correspondence *cp* is the usual plotting correspondence connecting a pair of numbers in the table with a point in the plane with those numbers as its respective coordinates. Sitting in the background are the two set-based descriptions of subsets of $N \times N$, the intensional $\{(X, Y) \in N \times N \mid Y = 2x + 3\}$, and the extensional, which lists a sample of elements of N × N. The former is also sometimes written as $\{(X, 2X + 3) \mid X \in N\}$.

Before turning to our lonely student and a host of syntactic details that lie in wait to stumble him, we note that Figure 14.4 is a diagram very different from those offered by Lesh, Goldin, and Janvier in PART 1, where each is attempting a description of representation acts in much more generality. Lesh, in particular, is also acknowledging differences in media within his diagram. Were we to locate the place of our Fig. 14.4, for example, in Lesh's general schematic, it would be within his circle described as "within symbols," because we include coordinate graphs among formal, synthetic symbols. However, the acts of learning and using such systems inevitably require the kinds of actions described in Lesh's diagram (i.e., sequences of speaking, reading, writing, etc.)

The calculations leading to the entries in the table and the entries themselves are in the Hindu–Arabic base ten placeholder symbol system. Depending on whether only whole numbers are used, the extension of that system to decimals or to fractional representations is also being employed, as are other extensions to include the arithmetic of signed numbers. Although we may speak of "numbers" in mathematical discussions, the description of someone using numbers, either instrumentally as in this case or as items participating more passively in defining some mathematical context, must include a description of how those numbers are being presented.

Notice that the symbol scheme *T*-2 needs further description: For example, is the table horizontal or vertical? A vertical table arrays the corresponding numbers in a way that parallels their order as coordinates in the plane, hence presumably requires at least one less cognitive step (a 90° rotation) in the plotting correspondence than does the horizontal table. A tacit piece of syntax in *T*-2 relates to the order in which (the symbols representing) numbers are listed. Unless otherwise directed, novices do not usually adopt this convention, nor are they likely to be systematic in other ways, e.g., in choosing numbers that constitute, say, an interval of integers.

Let us assume that the medium in which results are recorded is paper and pencil. The medium in which numerical calculations take place might be either the same pencil/paper or an electronic calculator. The permanence of marks generated affects the student's ability to interpose additional number pairs between those already written into the table—not an important matter here, but significant when a nonlinear function is involved—the paper/pencil medium does not support the kind of plasticity of the records that is often necessary (see Barclay, 1985). The impermanence of the intermediate marks generated by an electronic calculator, as opposed to those generated on paper during calculations, is likewise an issue. (Incidentally, the major initial factor guiding the choice of a linear function was the limitation of my computer and printer media regarding superscripts. Were I working with my new computer, I would have used a quadratic function with nonintegral vertex coordinates.) Each of these media issues, and indeed many others, are the subject of discussion and decision when designing a computer-based learning environment (Kaput, 1985b).

Let us now continue to the next part of the task, plotting the points whose coordinates appear in the table as adjacent numbers. Again, there are details to be specified, perhaps the most important of which is the syntactical issue of scaling and marking of the coordinate axes, which control the visual experience of the representation in conceptually important ways. Lesser details (such as whether the axes have arrows, are in different colors from the graph, etc.) can be subsumed in the specification of what kind of equivalence class of inscriptions constitutes a character.

Having sketched the line connecting the points thus plotted and having thus completed the task, the student has thereby established the correspondence between the symbol scheme *E*-2 and *G-2* in the symbol system (*E*-2, *G-2*, *cs*) by passing through the symbol system *T-2* (see Fig. 14.5). Note that the symbol *scheme E*-2 is participating in two different symbol *systems* in this task, (*E*-2,

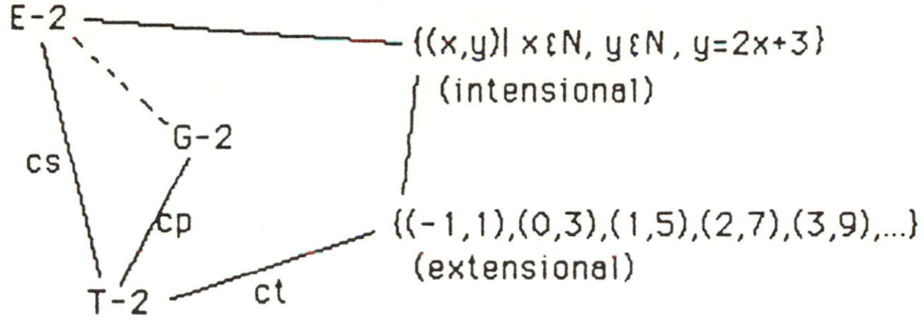

E-2 — $\{(x,y) \mid x \in N, y \in N, y=2x+3\}$
(intensional)
G-2
cs
csp
$\{(-1,1),(0,3),(1,5),(2,7),(3,9),\ldots\}$
(extensional)
T-2 ct

FIG. 14.5.

T-2, cs) and (*E*-2, *G-2, csp*). Here *csp* is the composite correspondence plotting the algebraic equation via substitution.

In an electronic medium, the plotting could be done automatically, either point by point with dots connected subsequently, or more quickly. As pointed out by Dugdale (1982) and in a more general context by Dickson (1985), the ability to plot graphs quickly and efficiently on a microcomputer can compress much more mathematical experience into a fixed amount of time than is possible in the paper/pencil medium, although most computer-based graphing experiences shortcut the use of *T*-2 by connecting *E*-2 directly with *G*-2. It is unclear what the costs may be of factoring out the numerical portion of the experience.

Tables and Graphs: A Symbol System Analysis

Let us now look at some more global issues connected with this example. Notice that the numbers involved in the task could have been chosen to be the integers instead of the real numbers, for example. But in that case there would be a formal failure of fit between the "discrete" $N \times N$-ten and *T*-2 on one hand and the symbol scheme *G*-2, which is "continuous" (although it could obviously be modified to represent discrete number pairs). On the other hand, the scheme *T*-2 is consistent with both number systems, but it represents a small sample of either. Moreover, the sampling process inherent in *T*-2 is pointwise (*T*-2 cannot represent intervals) whereas *G*-2 samples intervals. If *T*-2 were used to represent a linear relationship given as a proportion, say $3Y = 2X$, then the sampling process may be even more constrained (to contain only even values of X for example). Given differences in sampling, there follow differences in the kinds of inferences made. Because the symbol schemes are always in some imperfect correspondence with one another in very fundamental ways, and these ways are not often the subject of explicit instruction or attention, they can contribute significantly to student error phenomena.

Another source of difficulty relates to the choice of the symbols standing for the variables. For example, are they part (as axis labels) of the respective symbol schemes involved or not? This depends on how they are taught and learned and serves to explain certain student difficulty in using new characters in what the teacher may assume to be a symbol scheme that is independent of choice of character to represent "variable name." Moreover, symbol schemes are usually overlaid on one another—an equation often labels the graph, certain points are labeled, the table may contain some parts of the equation, etc.

More significant perhaps are questions about the relationships between certain features of the symbol systems and underlying mathematical concepts such as linearity in this case.

In *E*-2 the equation $Y = 2X + 3$ uses an efficient and direct shorthand to help us "see" a rule (not the only one, of course) that generates Y values given X values—a form of procedural knowledge not easily gleaned from representations in the other symbol system. It also makes cognitively accessible certain conceptual knowledge about the constancy of the quantitative relationship across allowable values of X and Y, although much is implicit here regarding the variation across the domain set. Indeed, the concept of algebraic variable is difficult in part because so much *is* implicit. However, it is important to keep in mind when discussing symbol system features such as these, that they do not reside "out there" in the structured characters themselves, but rather they reside in the interaction between the notation *and* a suitably learned user. As we argue shortly, the user only has access to his/her own interaction with the symbols—and does not have direct access to any external world, including a symbolic one.

In *T*-2 the quantitative relationship between X and Y is implicit whereas the variation (across a sample, of course) is explicit, assuming systematic variation of X across the domain. The implicitness of the relationship is not a factor in this task because the equation supplies it explicitly. In other tasks where the equation is not given, the inference of the pattern and its formal description (in *E*-2) can be a nontrivial exercise. The quantitative relationship can be made more explicit by expanding the symbol system *T*-2 to include "first differences," where the change in Y corresponding to a unit increase in X appears explicitly (it is, of course, the slope).

The case of *G*-2 is quite revealing. The graph provides a very efficient and explicit representation of BOTH the quantitative relationship and the variation. Why? First, the quantities involved are all automatically ordered. Second, they are simultaneously presented explicitly as points on the graph so that change in the relationship between the quantities (or lack of it in this case) can be readily seen. Third, they are much more completely represented than in *T*-2—they are more completely sampled. Fourth, instantiations of the quantitative relationship, namely ordered pairs, are compactly presented: Each such pair of numbers is consolidated into a single entity, namely a point. Hence to compare two pairs, one need only examine the relative position of two points rather than the relation among four numbers; and because of the first three characteristics of *G*-2 listed, many such comparisons can be facilitated. Fifth, one can apply much previously learned knowledge from visual experience, knowledge about straightness, slope, curves, crossing-points, parallelism, etc., to mathematical situations represented in *G*-2 (e.g., simultaneous linear equations).

Notice that the other representations related to this task, the set-based descriptions, possess virtually no qualities not already emobided in *E*-2, *T*-2, and *G*-2. The set-based descriptions, bare as they are, are useful mainly for attending to more abstract aspects of the situation, e.g., the common role of ordered pairs in the various representations.

Much more could be said about such matters, but they are mentioned to indicate that much detailed analysis is possible from the starting point offered here, and that many student behaviors and errors may have explanations in terms of a sufficiently rich theory of symbol systems.

Philosophical Preliminaries—Some Philosophical Engineering

The question of "referent" for a given mathematical character raises a whole host of philosophical questions, both new and old, relating to the ontological status of the referent as well as its relationships with thought. We must make a few comments both to avoid a barrage of premature philosophical criticism as well as to tack down some slippery issues. The author believes that symbol-use difficulties and the need for systematic understanding of representation systems are of sufficient importance to warrant treading in that treacherous no-man's-land where reside assertions about interactions between shared public symbols and cognitive processes. Armed with but the barest referential theory of meaning, clothed in a highly awkward hand-me-down objectivist language, we travel close to the edge of Platonism (Hawkins, 1985) and the brink of Nominalism, always in the mist of Psychologism. As just implied, our received language is deeply and inherently Platonist in its referential assumptions and objectivist syntactic structure. However, it is not our intention to cross the boundary into Platonism. In fact, as Hawkins notes (1985), even Plato was not really a Platonist. After all, even the light show in the cave required a medium and representational forms!

The uncomfortable reader should know that we are approaching the philosophical journey in much the same pragmatic way an engineer uses mathematics—function and utility are paramount; formal precision can wait, provided we are fundamentally consistent. Thus the phrase "philosophical engineering."

Shared Symbols

First of all, it seems entirely reasonable to assert that shared symbols exist and function at the collective level, and that their shared properties and referents can be the subject of systematic discussion. We make a working assumption that such shared symbols have a dual existence, one in the public domain and the other in the private cognitive domain. This duality reflects the dual, but intimately related, uses of mathematical symbols—one for communication and the other as an assist to cognitive processing. As public entities, the "correct" properties and usage of shared symbols are those that are defined by convention and agreed-upon formal fiat by the community of competent users. In some sense this is an empty statement, because " 'correct' properties and usage" can only be known by individuals, phenomenally, and do not exist "out there" apart from any knowers. However, for a variety of reasons it is useful to use this kind of objectivist language. As cognitive entities, symbols cannot be observed, and it is a matter of significant debate (e.g., Kolers & Smythe, 1979) how to speak of "symbols in the mind." However, we speak of the "cognitive version of the symbol system" and the shared or common features of the cognitive versions in a particular language community. It seems possible to do this without making too many premature commitments about their nature.

Relative to a symbol system used in a particular language community, the extent of cognitive "sharedness" of a given symbol's features, relations with other symbols, and correspondence with referents by a particular user depends on the cognitive makeup, hence experience, of that user, but is never ultimately knowable. It can only be inferred from careful, close, and extended observation. One way to express the aims and consequences of mathematics instruction for a learner is as the increase in the publically shared characteristics of symbol and referent that the learner can be said to have constructed or internalized.

Now the extent to which a population of 4-year-olds shares a cognitive referent (by which we certainly include an action) for the string "$3 + 2$" is likely to be less than, say, the extent to which a population of college students shares the cognitive version of the coordinate graph referent of "$y = 2x + 3$" in the symbol systems of two-variable algebraic equations and their graphs. By the time a person is dealing with the algebraic symbol system, enough mathematical experience has transpired to support the construction of many facets of the public symbol system and to delete obviously inconsistent features of individuals' cognitive versions. We can thus make more general and confident statements about the shared properties of the cognitive versions in the latter population than the former. In this sense the latter population of college mathematics students "shares" the symbol system in question more than the young children "share" the other symbol system. This sharedness results from an active testing of one's interpretations and reading of others' reactions, what we call "communication." Of course, without access to the contents of others' minds, we can never be certain regarding the *kinds* of features shared across individuals.

Stories of this sort, about communication and sharedness of symbols, do not, however, put to rest the deeper philosophical questions about reference or the existence of mathematical entities. Before moving further on these questions we do a bit more sifting of the concepts before us. In addition to the dual existence posited for symbol systems, private-cognitive and public-shared, we suggest that some mathematical systems play a more fundamental referential role than others.

Primary Reference Fields

What can be said about reference fields (in both modalities, shared and cognitive) such as the natural number system N (or $[N \times N]$, the field of binary quantitative relations on N as in the illustrative example)? These act as the field of reference for many other symbol systems, e.g., the base ten placeholder system.

We are asserting that, although symbol systems act as reference fields for one another, ultimately, chains of reference for symbol systems get back to the primary reference fields, such as the natural number system, which are taken to be mathematical structures. These primary referents are not other symbol systems *universally* expressible further in terms of shared symbols relative to a given language community, but rather entities with shared features in a given cultural community. We say "cultural community" rather than "language community" because such a structure seems to be readily shared across language communities—the way "3" is written or spoken seems unimportant to its meaning.

Also, we caution that by identifying mathematical structures as primary reference fields, we do not mean to imply that other reference fields do not embody mathematical structure. Indeed they do—it is that structure that provides the basis for the symbol systems built upon them. Finally, for the competent users in a particular language community we shall assume further that the cognitive versions of the *primary* referents are flexible, widely applicable cognitive entities with phenomenal stability and permanence. These can also be regarded as the phenomenal images referred to by Mason that serve as the constituents and pointers for the more complicated imagery that underlies the doing of more advanced mathematics.

The elements of primary mathematical structures often refer to "external" entities, including mundane symbols such as objects, if you will. They may also refer to "internally" generated entities such as the number of prime factors of 105. Thus we are making the working assumption that it makes little sense to speak of universal, externally sharable referents for the primary symbol systems. And, of course, we make the common assumption that all such cognitive referents of such a primary reference field as the natural numbers are built up from extended experiences with counting and with physical objects (the latter again acting as external, sharable symbols). For a fuller discussion of the problem of reference and its relation to cognitive models, see Kaput (1985a) and the discussion later in this section.

It will usually be the case, however, that in particular instances relating to a person doing a particular task, one can discuss sharable referents of, say, the character "3" in a primary symbol system, e.g., a picture with three ducks in a row, three actual ducks, or even the number of prime factors of 105. Indeed, this is where the construct of *quantity* helps make sense of the primary mathematical system's rich connections with nonmathematical referents. One must not lose sight of the fact, of course, that referents of "3" such as pictures of ducks, or "actual ducks," are shared referents *in the same sense as the shared symbols described earlier.*

Quantities and Numbers

A *quantity* can be defined formally to be an ordered pair (M, R), the first component of which is a number and the second of which is a referent for that number. The number describes how many or how much of the referent, e.g. (3, ducks). Quantities can be added, multiplied, divided, and so on, according to reasonable rules that reflect how numbers are in fact used in everyday life. These rules are reflections of the semantic relations among referents. For example, one can add 3 ducks and 2 chickens by first creating the superordinate reference set of, say, fowl, and then adding 3 fowl plus 2 fowl to get 5 fowl. For details regarding multiplication and division, especially "intensive quantities," see Kaput (1985b) and Schwartz (1984).

We do not develop a formal calculus of quantities here but simply note that (1) such can be defined (see Schwartz, 1976, 1984, and Schwartz and Kaput, in preparation), (2) it provides a systematic account of the role of the primary number symbol system's referential connections with the wider world of referents outside of mathematics proper, and (3) it provides a bridge to natural language from mathematical symbol systems. In the next section we discuss the role of natural language as an ambient representation system for mathematical symbol systems. Quantities can be thought of as playing a mediating role between the primary mathematical structure of (pure) numbers and the natural language.

Our Approach in Perspective

The preceding discussion does not pretend to answer ultimate questions of reference such as "to what does the character '3' refer?" Or "do sets and functions exist?" The author's view is that these are incorrect questions and need to be replaced by better ones. Such matters of reference are becoming secularized in much the same way that natural science and psychology have gradually moved from topics of speculative philosophy to topics of empirical study. This is not to say that there is no room for philosophy—indeed, analysis becomes all the more important, as it has, for example, in physics. The secularization of matters of reference and related questions of mathematical truth and ontology takes the form of the gradual intrusion of psycholinguistics, cognitive science, developmental psychology, the history and sociology of science and mathematics, cultural anthropology, and so on.

The move away from classic apriorism has been illustrated and argued well by such centrist authorities as Putnam (1975) among others, the move toward historicism is exemplified by Kitcher's well-received book (Kitcher, 1983), the cultural and social foundations of mathematical ideas and truth find initial exposition in the work of Wilder (1981), Kline (1972), and Bloor (1976), and psychologism has been creeping back through the crevices opened by these and other changes in philosophic perspective, e.g., in the work of Davis and Hersh (1980) among others. It is important to realize that all these reevaluations are functioning with extremely limited analytic tools and hence limited systematic data. But when such tools are in hand, such approaches will no longer be controversial but will represent mainstream areas of endeavor.

Thus, although it may be incomplete and unsatisfying to say, for example, that the reference buck stops at two such earthly places, the private cognitive level and the social level, and to speak of the ''sharedness'' of symbol systems in terms of learning, development, and socialization, these are but primitive pointers to the vast, mainly uncharted arenas of study that would eventually, as one segment of their achievement, account for the power of mathematics and its various symbol systems as intellectual and cultural instruments.

Lastly, the fact that mathematical abstractions, especially those that have the character of ''objects,'' such as sets and numbers, are *experienced* as real entities separate from oneself does not prevent us from regarding them as individual, personal constructions. In fact we see the heart of mathematical instruction to be in providing the opportunities, mainly actions and reflections on those actions, for building those phenomenological mathematical entities that, as Mason points out, become ''palpable'' as they are used. Ultimately, for individuals they are the entities to which shared symbols refer.

Skepticism, Encodingism, and Frames of Reference

One risk of our—and indeed any—description of representation that posits two sides and a correspondence between them is what Bickhard calls the ''incoherence of the encoding approach'' (Bickhard & Campbell, 1985), seen especially clearly if one wishes to pursue the idea of cognitive representation. The problem is roughly this. If one's conception of a representation act assumes that the representing world Y must *encode* information present in the represented world X, then the representing side Y must already have the information or ''know'' what is being represented. This general observation has consequences at two different levels at which the word ''representation'' is commonly used: (1) at the global cognitive representation level, where X is the ''external world'' and Y is the mental representation of that world; and (2) the local symbol system level, where X is a piece of mathematics in one symbol system and Y is a representation, or re-encoding, of that mathematics in another symbol system. We deal with each of these situations in turn before turning to an interaction between them.

The author's position is that direct knowledge of an external world is an epistemic impossibility—hence cognitive representation of an external world makes no sense. On the other hand, one does interact with an external environment. These interactions produce mental events that are in turn acted upon by the mind to produce what we and others are in the habit of calling mental (or more narrowly) cognitive representations. Thus the interactions produce internal mental events that constitute X the represented world, and the actions on them construct Y the representing world. No *knowledge* of an external world is involved, only interactions with one. See von Glasersfeld's lead-off contribution to this book and his references for a greatly refined extension of this position.

This situation is fundamentally different from the use of a symbol system, say the two-dimensional coordinate graphical system, to represent a linear equation, $y = 2x + 3$. Here the "knower" HAS access to both sides of the representation (although he or she may not understand both sides or the correspondence equally well). Thus in this case the idea of encoding makes epistemological sense. Moreover, to understand the linear relationship being encoded in both forms is a new understanding that is encoded in neither form, but rather in the entire ensemble, including and especially the correspondence between them. (See Lesh's first part in this volume for another version of this assertion, and see Kaput, 1985b, in press, for suggestions on how computers may be used to make multiple representations more learnable through linked simultaneous presentation.)

Therefore, the two uses of the general idea of representation are epistemologically different. But there is a third, frame-of-reference, issue that relates both of them to our notion of "shared symbol systems." We have noted the direct epistemological inaccessibility of the environment to individuals. But for this chapter, one key feature of the environment is the set of cultural artifacts constituting our inherited mathematical symbol systems. How, then, is it possible to discuss something that is inaccessible and, even further, to discuss the extent to which it is "shared" by individuals? The answer is provided by the realization that observers of an individual *do* have access to (their own constructions of) the individual's environment and interactions with it. Thus the observer has a form of access that the individual does not have, and, on the basis of observation, can thus make claims regarding how that individual interacts with the inherited mathematical symbol systems, for example. Such an observer can likewise make legitimate inferences regarding the extent to which different individuals "share" a particular symbol system on the basis of their exhibiting similar behavior with respect to that symbol system.

This is, of course, a relatively weak form of sharedness because we have no privileged access to the contents of anyone's mind, and our sharedness assertions are about individuals' internal representatoions of their interactions with the symbol environment (their phenomenal mathematical entities) and not about their representations of that environment—which are epistemically impossible.

However, it is the only form of sharedness possible outside of the solipsist's mirrored isolation chamber.

Two Types of Symbol Use in Mathematics

We have already referred to the dual existence and application of symbol systems, public, communication, and private, cognition. (Note that we are in agreement with von Glasersfeld (Part 1) in subscribing to the constructive rather than the transfer metaphor for communication.) Cognitive activity associated with symbol systems itself splits into two types, often present in complex interactions (Salomon, 1979).

1. Reading and Encoding. When confronted with a set of symbols one must read or interpret them, or when one is confronted with a task requiring the use of a particular symbol system, one must encode information into that system from some other symbol system. These are both cognitive activities that require the application of previously learned skill and anticipatory frames. Usually the skill is comprised of a heirarchy of subskills associated with the different levels of processing required to interpret the given symbols. The best reading example is perhaps provided by the most completely studied such skill, the reading of natural language, where the chunking of elemental processes into automatic subroutines of larger processes (e.g., letter and phoneme recognition) are, with practice, chunked into word recognition subroutines, etc. We examine a variety of encoding situations later in this chapter.

2. Elaboration. After having read a symbol or having encoded information into a particular symbol system, one uses the symbol system in one of two ways: (1) *syntactic elaboration,* by manipulating directly the symbols that represent the entity involved, either mentally without written help—as when one does mental arithmetic, say the addition of a pair of two-digit numbers, by thinking in terms of the symbols themselves (as if on a mental scratchpad)—or with written or other extracortical help, e.g., on a "real" scratchpad; (2) *semantic elaboration,* by elaborating referents of the symbols, as when one solves the simple linear equation $X + 3 = 7$ not by mentally subtracting 3 from both sides in a mental version of the equation, but by accessing one's knowledge of whole-number addition, perhaps by sequential trial and error, to determine what number added to 3 yields 7. Here the elaboration is done by using the features of the reference field of the symbol system rather than by using its symbol scheme syntax (which includes rules for transformation of equations into equivalent equations).

The two distinctions jusi made are not meant to imply clean or simple divisions of labor or labor types. In fact, reading/encoding and elaboration intermix in complex ways. One such example is a "visually moderated sequence" (Davis, Jockusch, & McKnight, 1978), where a person cycles between writing and reading symbols, with the output from each stage of the process read as the determinant of the next step in the process. Familiar visually moderated sequences are those involved in arithmetic or algebraic long division.

In Part 1 Mason describes attention "on the symbols" versus attention "through the symbols," a distinction entirely parallel with what we are trying to capture here in somewhat more formal terms. Skemp (1982) likewise speaks of surface versus deep structure. Similar parallelism with Mason's position occurs when we describe how the use of names for mathematical structure or processes helps reify them into phenomenal entities that can then participate in other structures or processes.

The other distinction, between syntactic and semantic elaboration, is also complex on two grounds: First, they are but polar extremes that shade into one another in intermediate positions; second, a particular individual's approach to a task will likely slide from one type toward the other over time, depending on intervening instruction and/or practice (e.g., Baroody, 1985). Indeed, there seems to be a natural progression from semantic toward syntactic elaboration as the actions on symbols, or relationships constructed among them are reified into entities that can in turn serve as referents for new symbol systems (more on this later).

Often, the semantic elaboration takes the form of actions with relatively primitive, even sensorimotor, roots—fractions as action-based operators, functions as movers or transformers of numbers, etc.—before the actions yield entities that can serve as referents for another symbol system—fractions as points on a number line, functions as objects that can be added or differentiated, etc.

Although one may view these observations as a restatement of Piaget's idea of reflective abstraction, it seems that Piagetians have yet to acknowledge fully the role of symbols in this process.

These distinctions are thus first intended to assist systematic discussion of particular symbol-use phenomena. This discussion in turn will lead to a fuller theoretical development of those elementary distinctions. One mark of their value will be the extent to which they unify previously unrelated phenomena and provide the means for plausible, testable generalizations about symbol use in mathematics.

Knowing: Syntactic–Procedural Versus Semantic–Conceptual

The symbol system perspective may provide a new angle on the long-standing procedural/conceptual distinction. The basic assertion is thus: Procedural knowledge is based much more on the direct manipulation of immediate symbolic representations than is conceptual knowledge. Let us take a brief look at some common situations.

Arithmetic Computations. Carraher and Schliemann (1985) examined the solution processes and associated error patterns on a set of computation problems using two categories developed by Reed and Lave (1981): those that were based on actual quantities (including finger counting and the like), and those "in which the manipulation of symbols carries the burden of the computation" (Carraher & Schliemann, 1985, p. 37). For example, "a" types of procedures for the subtraction problem "21 − 8" include counting down, counting up, or decomposing the 21 into a 10 and 11 and subtracting the 8 from one or the other, etc. The most likely "b" solution was via the borrowing algorithm, which involves actions on the symbolic representation of the problem itself. In Bruner's terms (1973) these are "opaque" uses of the symbols, rather than the "transparent" solutions, whereby the user "sees through" the symbol scheme to be guided by its field of reference.

From our perspective, the "a" solutions were semantic elaborations based on using understood relationships among the elements of the reference field N for the symbol system (N-ten, N, c). The "b" solutions were syntactic elaborations based on the syntax of this symbol system's symbol scheme. The observed errors for the two types were vastly different, with the most salient difference being the lack of quantitative reasonableness of the syntax-based errors (e.g., subtraction errors where the answer to "$x - y$" was larger than x for $0 < x < y$). A similar dichotomy of solution procedures (syntactic vs. semantic) occurs in word problems when the reference field is a real-world situation. In each type of problem situation, the two classes of phenomena at hand are directly describable in terms of the structures of the domain in which the solution procedure is taking place—at either end of the correspondence c between a symbol scheme S and its reference field F.

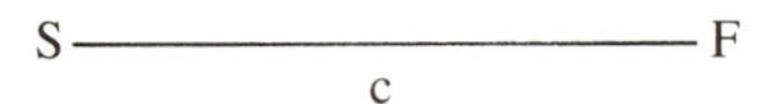

Of course, in most cases the reference field will be another symbol system rather than a primary mathematical system as in the previous case, and often, as with rational numbers or calculus, several interacting mathematical ideas and representations would be involved.

Algebraic Equations. We have already noted how a simple linear equation in one variable can be solved using knowledge of the numerical reference field for the algebraic symbol system. For example, prealgebra students solve $2X + 1 = 7$ by answering a question something like "If we double a number and then add 1 to get 7, what is the number?" For such students, the problem, whether or not solved by using mental operations paralleling a syntactic solution (subtract 1 from both sides and divide both sides by 2 to isolate X), is approached via elaboration in the numerical reference field for the algebraic statement. On the other hand, the solution to $\frac{2}{3} - 4X + 81 = 11 - \frac{3}{2}X$ is not likely to be solved via a semantic elaboration in a quantitative reference field, but rather using syntactical rules (some of which involve arithmetic rules to be sure) operating on the symbols that comprise the equation itself. Of course, the essence of the system's power is in the rules that, if appropriately executed, preserve the quantitative relation expressed by the original equation. Indeed, the value of the system is in its ability to empower its user to solve equations that he or she cannot comprehend numerically.

In the same way, the power of the multiplication algorithm in the base ten placeholder system is embedded in the fact that the symbol manipulation faithfully preserves actions on the referents of the symbols representing products that are not comprehensible in and of themselves—the user can trust the algorithm to produce a symbol that corresponds to the answer. To appreciate the last statement, imagine a multiplication algorithm that, instead of sliding the result of multiplication bv the tens digit one digit to the left of that for the units digit, recorded the tens product one digit to the right. By following such a rule and following all the other rules as usual for 23 times 56, one would get the symbol "1792" instead of the correct symbol "1288."

Another example of syntactical symbol use occurs in the solution of simultaneous linear equations by either matrix or determinant methods (Cramers Rule), where one manipulates coefficients taken from the system's representation without regard to their various meanings, which are incorporated into the various algorithms involved. Solutions by substitution are closer to semantic applications because of their use of the conditions that define the equalities involved.

Precalculus and Calculus. Similarly, a precalculus student probably could solve $\log X = 0$ by using knowledge of the quantitative relationship represented by the equation (including the definition of the logarithm function: "*e* to the zero power equals what?")—a solution by semantic elaboration. However, to solve $\log (X^2 - 3X + 2) - \log (X - 1) = 7.32$ one needs to act on the symbols themselves, again according to certain rules—a syntactic elaboration. Of course the simpler equation could also be handled syntactically (take "antilog" of both sides).

In the same vein, one can know the derivative of $f(X) = 2X + 3$ either by applying a derivative rule, or with respect to a reference field for the function, say the graphical reference field examined earlier. In the latter case the derivative would be known in terms of the slope of the graph. Once again, however, it is easy to provide examples where the derivative can only be determined through the application of a rule on the symbols representing the function.

In each case we have alluded to, one sees an enormous increase in power resulting from the development of a symbol system whose scheme's syntax and correspondence faithfully preserve certain relationships in its reference field. As elaborated a bit in the next section on the tool aspect of symbol use, the building of new symbol systems is among the key ways that mathematics evolves, both historically and within individuals. For individuals however, the power and freedom of the formalism brings an added responsibility—the ability to interpret the formal procedures back in the reference field(s). Otherwise, all one has are meaningless symbols. Meaning exists relative to the appropriate reference field. (Again, ours is a referential theory of mathematical semantics.)

But given enough experience with the new symbols, they, their relations, and actions upon them too become stable phenomenological entities and hence the referential base for yet another symbol system. As evident in the next section, such a process varies tremendously from one type of mathematical symbol system to another, depending on the mathematics, the aspects being emphasized, and the type(s) of symbols being used.

More on Symbol Systems as Tools of Thought

It is well known that a particular representation can unlock a problem situation or provide a particularly powerful tool to think with in particular instances (Olson, 1985). We have hinted at the outset of this chapter and in the preceding discussion how a whole new set of actions and applications of a mathematical system becomes possible with an appropriate symbolic representation, as with the modern number system. It is also worth indicating how the historical growth of mathematics itself is influenced by the invention of new representations.

Whereas a symbol system can be used to represent a mathematical structure, a symbol system can also give rise to new mathematical structures by providing the means by which one can think more abstractly—as is the case with algebraic symbol systems that support the abstraction of properties from particular mathematical structures to create new ones, such as the various classes of algebraic objects (groups, rings. fields, etc.) The essence of this function seems to lie in our ability to capture or highlight certain features of a mathematical system and suppress others by varying the features of the symbol scheme representing it. For more examples see Chapter 2.

Another closely related function of symbol use is based on applying the syntactical properties of a given symbol system's symbol scheme to a new field of reference, as with the definition of metric spaces or normed linear spaces, where the basic structures captured by familiar metrics on familiar cartesian products of numbers are applied to new classes of objects such as sets of functions or even operators. More mundane versions of this phenomenon occur as a number system expands, e.g., to include negative or rational exponents.

Indeed, the development of the complex numbers provides several instances of notational advances leading first to increased computational power based on the syntax of the system (of real numbers) being extended, and then to an acceptance of the reality of the extended system. Bombelli's 16th-century idea to admit roots of negative numbers into his calculations with roots of polynomials allowed for computations that related these newly recognized roots of polynomials to real numbers. Burton (1985) quotes Bombelli's reflection on his rather cavalier extension of real number notation: "It was a wild thought in the judgement of many; and I too for a long time was of the same opinion. The whole matter seemed to rest on sophistry rather than on truth" (p. 314).

Later, Gauss (and in fact others) invented the two-dimensional plane representation of complex numbers—although even at that time the letter "i" (Euler's choice) was not in common currency. Nonetheless, the "imaginary" numbers could be seen and operated with. The status of complex numbers as phenomenal mathematical entities for individuals was enhanced. As a result they became more "real" and acceptable, eventually becoming a shared mathematical symbol system among the community of mathematicians.

About 300 years after Bombelli, in 1849, Augustus de Morgan wrote in his text *Trigonometry and Double Algebra* (quoted in Cajori, 1929b):

> As soon as the idea of acquiring symbols and laws of combination, without giving meaning, has become familiar, the student has the notion of what I will call a *symbolic calculus;* which, with certain symbols and laws of combination, is *symbolic algebra;* an art, not a science; and an apparently useless art, except that it might afterwards furnish the grammar of a science. The proficient in a symbolic calculus would naturally demand a supply of meaning. (pp. 130–131)

Presumably de Morgan was writing about algebra, but his comments apply much more widely. The geometric representation of complex numbers helped provide the meaning, another field of reference, for the complex number symbol scheme.

An exactly parallel situation has occurred in Chapter 15, where Lesh's addition—using the SAM system—of a graphical meaning to operations on equations now provides a "visual semantics," that is, a new graphical field of reference for actions applied to equations. (A similar system has been developed by Michal Yerushalmy and Judah Schwartz.) Previously, one in fact operated on equations, but it was seldom described in these terms.

There are also symbol systems whose functional features extend to be used much more broadly than their original application. Descartes coordinate plane invention came to be used as a general tool and hence provided the basis for a field of reference for a wide variety of algebraic and geometric systems of virtually any dimension.

Our emphasis on the role of symbols in the development and use of mathematics should not be misinterpreted, however. To assert the identity of mathematical structures and symbol systems is very close to asserting the formalist philosophy of mathematics—mathematics is defined by the manipulation of symbols according to certain explicit rules. This certainly is not the position put forward here.

Indeed, the relations between mathematical structures and symbol systems are much more complex, involving two-way interactions that vary greatly from one system to another, both mathematical and symbolic. Nor are we espousing a nominalist position, that there are only names.

Equivalence of Representations: Syntactic and Semantic

That symbol systems differ is obvious, but characterizing HOW they differ in a way that is logically consistent, psychologically significant, and mathematically relevant is not a simple matter. We continue laying a theoretical framework to help build toward this goal by specifying some of the ways representations can be the same or different. The key is to respect the different levels of organization of meaning embodied in symbol use. Before doing this we recall a critical idea from mathematics.

Equivalence as formally defined in mathematics applies to elements of sets. It is defined as a relation satisfying three conditions: reflexivity (any element should be equivalent to itself), symmetry (the equivalence of two elements should not depend on the order in which the equivalence is stated), and transitivity (a first element equivalent to a second, and the second equivalent to a third, should imply that the first is equivalent to the third). These three conditions are necessary and sufficient to guarantee that any set with an equivalence relation defined on it is automatically decomposable into disjoint equivalence classes comprised of mutually equivalent elements, and conversely, any such decomposition of a set into disjoint subsets gives rise to an equivalence relation whose equivalence classes are exactly those subsets. (See any elementary abstract algebra text for details.) This is an extremely powerful and general idea that we now incorporate into our theory of mathematical symbol systems. (In fact we have already mentioned it earlier when we noted that a character in a symbol scheme could be defined as an equivalence class of inscriptions.)

Somehow, we need to specify how the two equations

$$Y = 2X + 3 \text{ and } U = 2V + 3 \qquad (1)$$

are the "same," and how the three equations (in E-2)

$$Y = 2X + 3 \quad 2Y = 4X + 6 \text{ and } Y - 3 = 2X \qquad (2)$$

are the "same." Clearly, different criteria should apply. Or in arithmetic,

$$2{+}3 \text{ and } 2 + 3 \qquad (1)$$

are similar in a way different from

$$2 + 3 \text{ and } 4 + 1 \qquad \text{or } 2/4 \text{ and } 3/6. \qquad (2)$$

Our first level of refinement of the idea of symbol system equivalence is embodied in the following pair of defintions, where in each case the equivalence is assumed to be a true equivalence relation on the elements of S.

Two symbols in (S, F, c) are *semantically equivalent* if they correspond to the same element of the reference field, i.e., $s \sim s'$ in S semantically if $c(s) = c(s')$ in F.

Two symbols in (S, F, c) are *syntactically equivalent* if their equivalence can be defined soley in terms of the symbol scheme S and its syntactical rules, and if the syntactic equivalence implies semantic equivalence. This second condition should in fact be incorporated into a more formal definition of symbol system—after all, of what use is a symbol system where, for example, two strings of symbols that can be transformed into one another using purely syntactic rules have distinct referents?

By these criteria, and under standard syntactic conventions, the symbols labeled (1) preceding are syntactically equivalent within their respective schemes, provided we specify the semantic equivalence.

On the other hand, to specify semantic equivalence in the items labeled (2) we need to be more specific regarding the reference fields and correspondences involved. Let us assume that the reference field for the equation scheme E-2 (over some fixed set of numbers N, say the real numbers) is $[N \times N]$, the set of all subsets of $N \times N$ where a specific subset is represented as a set of ordered pairs in intensional notation. Then we have duplicated the standard conditional equivalence of equations if the correspondence is the usual one, namely associating an equation with the set of pairs satisfying it—provided we ignore the differences in the representations of the set of ordered pairs!

Usually, as in the previous case, the reference field is itself a symbol system. This suggests the following slightly weaker version of semantic equivalence:

Two symbols $s1$ and $s2$ from S in (S', F', c'), c)are *semantically equivalent* if they correspond to syntactically equivalent elements of the reference field, i.e., $c(s1)$ and $c(s2)$ are syntactically equivalent in S'.

This now allows us to use the rules of S' to decide which referents are equivalent and thereby get around the difficulty of the previous example—where there are many different, but trivially equivalent, ways of expressing a set of ordered pairs.

In a graphical situation we would likely want two graphs of conditionally equivalent equations that are scaled differently to be syntactically equivalent, at least if the scales involved were all linear. One point of these remarks regarding graphs is that semantic equivalence is always defined with respect to a specified reference field, so that if the "meaning" (referents) of the elements of a symbol scheme change, then so may also the associated meanings of "sameness." (There is no change in this case, however.) A second point is that a variety of judgments must be made in applying the informal definitions given because of our informality regarding syntactical rules. Such judgments can help clarify one's operating or teaching assumptions on a given system or assist in the analysis or building of a new one, especially in the computer medium, such as the "SAM" utility described in Chapter 15 by Lesh, or the *Geometric Supposer* (Schwartz & Yerushalmy, 1985)—see later.

Reflections on the Definitions

The equivalence definitions generalize Goodman's (1976) dimension of "notationality," which requires that the correspondence be one-to-one (and the characters in the symbol scheme be discrete). The standard base ten placeholder symbol system *N-ten* = (*N*-ten, *N*, *c*) is notational over the counting numbers provided that the rules for combining symbols do not include the arithmetic operations. Notationality is lost, however, even for the usual extention of this system to include the negative integers, where the symbol scheme *N*-ten is extended to Z-ten via the usual emendation of *N*-ten by the "-" character. No longer is the correspondence 1-1 ("- -3" and "3" correspond to the same referent).

Our definition of semantic equivalence was not intended to capture the notion of variable from which algebra draws its immense economy and power. A given quantitative relationship over some set of numbers *N*, represented by an algebraic equation in two variables, e.g., our old friend $Y = 2X + 3$, is a subset of $N \times N$. The fact that the single equation (with an implicit choice of domain, of course) represents the entire subset is its source of economy.

Another aspect of semantic equivalence not explored in this chapter involves what may be described as a distinction between "term equivalence" and "conditional equivalence." An algebraic equation may be regarded as a statement of conditional equivalence between either functions, sets of numbers, or other entities, depending on the context and conventions in force. Then the actions on equations are either equivalence preserving or not, depending on whether the solution set or some other significant feature identified by the equation is invariant. On the other hand, one may perform actions on one side, say combining like terms, without involving the conditional equivalence, but only involving what we call term equivalence. Similar uses of term equivalence occur in the manipulations involving the individual parts (numerator or denominator) of fractions—rather than simultaneous action on both parts of a fraction. This topic will be explored further in a subsequent paper.

Before closing this part of the chapter we should point out that we have but implicitly dealt with a major underlying issue that is exposed if we rethink the definition of symbol system as a triple of entities (*S*, *F*, *c*) given at the outset. If *c* is a function from *S* onto *F*, then we have asserted that two symbols are semantically equivalent if *c* maps them to the same element of *F*. Why isn't ANY functional relationship between two sets inherently a symbol system with an automatic semantic equivalence resulting? (Such results by virtue of the fact that the preimages for any function are equivalence classes for the equivalence relation defined by declaring two elements equivalent if they get mapped to the same element of the range.)

This is fundamentally the point made in our earlier chapter—that mathematics is inherently representational in the way it operates. However, the companion point is that such a view leaves out the actuality of symbol-use phenomena. Not every function defines a symbol system in a practical sense. This said, we have not given an explicit set of criteria that would discriminate between those that do and those that don't. Nor can we—at least at this point in our development. We must rely on a tacit understanding based on the examples employed.

SECTION B: INTERACTIONS WITH NATURAL LANGUAGE AND PICTORIAL SYSTEMS

Introduction

No introduction to mathematical symbol systems can ignore interactions with the "natural" representation systems of natural language and pictures. These systems have a privileged status in our analysis because they are learned beginning virtually from birth and are thus the means bv which most other symbol systems are interpreted. In effect they act as the "ambient" symbol systems in which all others are embedded in the following sense: (1) The cognitive processes used to read symbol systems will default to those automatic processes engaged by these systems unless some kind of intervention takes place, triggered either by explicit cue or by tacit contextual cues; (2) these systems, especially natural language, are often used in combination with other special purpose or synthetic symbol systems. Virtually every mathematical definition is a mix of natural language and special mathematical symbols, and even the pure symbolic definitions are embedded in natural language text that serves as the real source of explanation—including explanation of how the mathematical symbol system works, i.e., explanation of its syntax and rules of correspondence with the reference field. Even the *Principia* has a natural language introduction.

Rather than attempt to build a systematic theory of interactions, which, given the complexity of the phenomena involved would be premature at best, we offer a few samples of interactions marked by conspicuous and well-studied effects at or near the heart of the mainstream curriculum in algebra, geometry, and beginning calculus. We are aware of many others, especially in arithmetic and will explore these in a later paper, but have chosen to concentrate on a few here that seem to embody some very general characteristics.

The Default to Natural Language Encoding

Many errors are based on the automatic "default" nature of natural language encoding processes. In particular, we trace the common sources of the meanings of the mathematical and natural language symbols that lie behind this default tendency. It is in these terms that we are able to discuss much of the data on the well-known Students–Professors Problem: At a certain university, for every 6 students there is 1 professor. Let S stand for the number of students and P stand for the number of professors. Write an algebraic equation using S and P that gives the relation between the number of students and the number of professors at that university.

The consistently high error rates (40–80%) on this task across age and ability levels, and the predominance of the "6S = P" reversal error (roughly 65% of all errors) indicate a difficulty in encoding the given information into the required algebraic symbol system, with the central source of the difficulty being the override of algebraic rules of syntax and rules of reference by those of natural language (Clement, 1982; Clement, Lochhead, & Monk, 1981). This and other data (Kaput & Sims-Knight, 1983; Sims-Knight & Kaput, 1983) indicate strongly that the associated quantitative relationship is well encoded (understood) in natural language and numerical terms but that the difficulty is at the initial encoding level.

For example, first-year algebra students who had little experience in writing equations produced fewer clean reversal errors than more experienced students but made up for the difference with other fragments involving the symbol "*6S*." In effect, they were less skilled at the elaboration aspect of the task, but their encoding errors were similar to those of more experienced students. When the context for the given numerical relationship was more familiar (involving, say, exchanging five pennies for a nickle), the errors involving *6S*-types of phrases increased significantly over those cases where it is arbitrary (five assemblers for each solderer in a factory). This indicates that the likelihood of shortcircuiting the algebra encoding is even greater when the natural language phrasing is prelearned and available to use without significant preliminary formulation.

The teaching experiment data from Wollman (1983) and other data involving mathematically mature subjects indicate that the natural language encoding difficulties are present even in sophisticated subjects, and that those producing correct equations often need to begin the task by consciously overcoming the natural language "default" encoding process that leads to the error. Lastly, it should be mentioned that the other side of this analysis involves the students' weak understanding of the notion of algebraic variable and hence the weak alternative to the strongly learned natural language encoding process to which it falls victim.

Similar analyses apply to other similar tasks and strikingly similar data involving translation into the algebraic symbol system from diagrams or from tables of data, but with additional mediating steps whereby the given quantitative relationship is first translated into natural language before the erroneous equations are produced. This also emphasizes the role of natural language as an "ambient" symbol system mentioned earlier.

Natural Language and Mathematical Symbol Systems: Common Roots

But of course the ambient and the synthetic systems of representation are not related simply by contiguity—many aspects of the synthetic systems have developmental roots in natural language, and their natural roots continue to feed the interpretation of the synthetic systems henceforth. We have already noted the evolution of numbers, from adjectival to nominal status, and we have likewise noted the origins of multiplication of numbers in repeated addition, which again takes its initial meaning from natural language: given six sets of objects, the total number of objects (it is no accident that the addition-related word "total" is used here) is obtained by adding the respective numbers of objects in each set.

These three sources of conceptual structures are thus integrated, mainly by superposition, when numbers are used in both ways in combination with multiplication. Consider, for example, how the numbers are being used when one speaks of "6 elevens" and writes $6(11) = 11 + 11 + 11 + 11 + 11 + 11$. The "11" is functioning in a nominal way, as an object in a set of numbers with its adjectival status now hidden, because we have suppressed the matter of "11 of what?" The "6" is both an adjective, telling us how many 11's there are, *and* a multiplier, yielding the product 66.

This composite interpretation is then extended to algebra, where the "6" can now function as a coefficient of a literal, say X, which stands for a variable. As before, the primitive adjectival meaning of "6" is preserved, as is the primitive repeated-addition meaning of multiplication. Hence now we see algebraic multiplication introduced via equations of the following sort: $6X = X + X + X + X + X + X$.

Because the X is usually introduced as standing for an "unknown," a single number, the tie with the "6(11)" equation is all the more direct.

Another tie with natural language developing in parallel is the use of adjectival numbers in association with quantities with formal units usually abbreviated to a single letter, e.g., "6*g*" for "6 grams." This use of numbers in extensive quantities is different from the use of intensive quantities (e.g., "6 students per professor," or "3 feet per yard") that supports a different model of multiplication (see Kaput, 1985b).

Finally, the "=" character and the word "equal" are likewise growing in meaning from within mathematics and in natural language. Within mathematics it has already the meanings (among many others to be sure) expressed in the previous two equations. It also has an equivalence-of-measures meaning that is tied to its use in unit-conversion quantity statements that take the form of "3 feet = 1 yard." Notice that if x stands for the number of feet in a particular measure and y stands for the number of yards, then the *algebraic* representation of the measure equivalence is "$3y = x$."

Putting all these crossed strands of meaning together, it is easy to see why the error rates on the Students–Professors Problem are so high and why the errors tend to take the forms that they do. More generally, however, we can see that with such deep-seated overlaps of meaings, the default routes connecting the mathematical and the natural language interpretations are many and well established—not only in this case but quite often across mathematics.

Pictorial Default Processes

The tacit automatic processing assumptions by which pictures are encoded parallel natural language encoding errors. The fundamental processing feature of picture reading is its automaticity, which tends to shortcut further processing and elaboration—a tendency to conclude without sufficient processing that what you see directly represents what there is. As with natural language, its roots date virtually from birth, are likely the result of strong evolutionary selection, and its effects need to be consciously overcome in many mathematical tasks involving picture-like representations, chief among which is the reading of graphs. It is in this vein that one can discuss the misinterpretation of graphs described by Janvier (1978), and Clement (1985) and which have been replicated in our own work (e.g., Sims-Knight & Kaput, 1983). One such task is to describe the shape of a racetrack given the velocity graph of cars doing a lap around the track (see Fig. 14.6).

The overwhelmingly frequent response is to give (or choose from alternatives provided) a picture that resembles the shape of the velocity graph rather than a more appropriate shape as in Fig. 14.7. This error is but one example of the class of errors that reveals an inability to distinguish between a function and its rate of change (derivative) found in our data and in that of many others, e.g., (Hitch, Beveridge, Avons, & Hickman, 1983). Another common error of this type occurs when students are asked to draw or select a graph of the velocity of a cyclist traveling over a hill—either with a hill pictured or described verbally. Typically, the preferred choice is a hill-shaped graph, rather than a more appropriate valley-hill-shaped graph.

When the effects of natural language and pictorial processing are combined, the consequences for student ability to encode a quantitative relationship into algebra are devastating. When given a Students–Professors type of problem with an accompanying schematic picture as given in Fig. 14.8, the percentage of

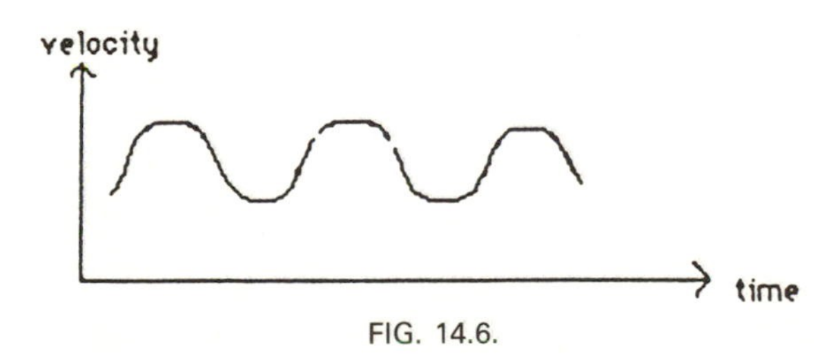

FIG. 14.6.

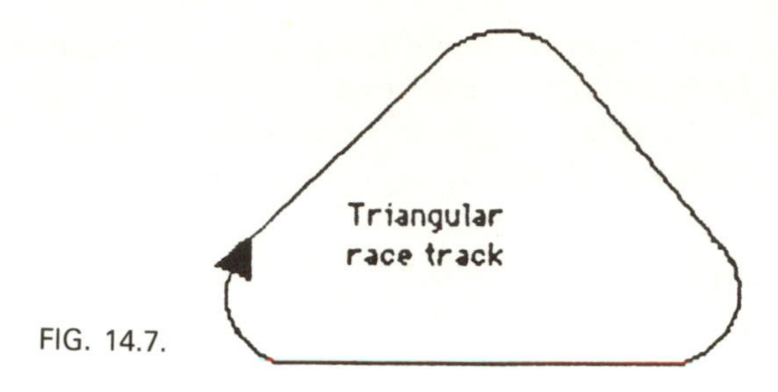

FIG. 14.7.

correct responses for a sample of 168 calculus engineering students fell from 60 to 37% for the half who were given the picture accompaniment.

Interestingly, the increase in errors was not a simple increase in the percentage of reverals but rather was based on an increase in a wide pattern of errors involving a "*6S*" type of phrase, a pattern of errors that was present in significantly higher percentages for other tasks with imagable referents for the given quantities than for nonimagable or linguistically supported quantitative relationships ("five pennies for one nickle") (Sims-Knight & Kaput, 1983).

Interference of Previous Syntax or Referents With New Referents or Syntax

It is reasonable to assume that the cognitive mechanisms underlying interference of deeply learned natural representation systems with the learning and use of mathematical symbol systems will be similar to those mechanisms underlying interference of a previously learned mathematical symbol system with the learning and use of a new mathematical symbol system. Usually either the old rules operate under new constraints or the old (usually tacit) rules no longer apply.

In this category of phenomena are some of the well-known difficulties and errors in representing rational numbers, and the difficulties experienced by students in ordering decimals (Nesher & Peled, 1984). The tacit rules learned from the base ten placeholder symbol system for whole numbers relating the length of the sequence of digits in the representation to the size of the referent do not

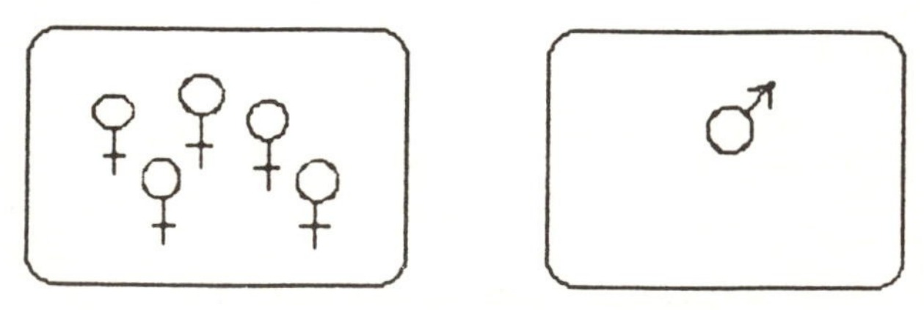

FIG. 14.8.

extend to the augmented symbol system for rational numbers. Similarly, Nesher and Peled (1984) have data showing that certain previously learned ordering features of the standard fractional symbol system interfere with ordering of decimals—for example, some students will decide that .42 < .23 "because fourths are smaller than halves." This category of phenomena also includes some of the familiar overgeneralization-of-linearity errors that occur in algebra and arithmetic (Matz, 1980).

Mixing and Interaction of Different Symbol Systems

Another important class of phenomena arises in connection with the coordination or mixing of symbol systems that are different in important ways. These difficulties occur whenever one labels a picture or diagram (which is a type of picture) using algebraic variables. Here two different symbol systems with very different assumptions are being mixed.

For example, in the standard first calculus course, the most common introduction to optimization problems uses the "Fence Problem" (maximize the rectangular area enclosed by a given perimeter as in Fig. 14.9). The standard approach to such problems is to sketch a representation of a rectangular region and then label it using variables. The intent is to represent thereby a "variable" region. The difficulty faced by many students, however, is that they see a single region, whose dimensions happen to be unknown—the fixedness of the pictorial representation is in direct opposition to the variation implicit in the algebraic representation. Given the weakness in student understanding of variable alluded to earlier, the picture helps "freeze" or limit the student's assimilation of the problem situation to a "solve for the 'unknown'" problem, one for which application of the differential calculus makes no real sense.

Because the labels for the picture are then used to set up the function that is to be maximized using a derivative, the derivative is being applied to a function without a true variable. It should be no surprise that the majority of students can make no sense of such tasks.

Although this misinterpretation is rooted in an inadequate conception of variables, it is also promoted by a tacit assumption associated with the reading of pictorial symbols—the assumption that a picture represents a single "thing"

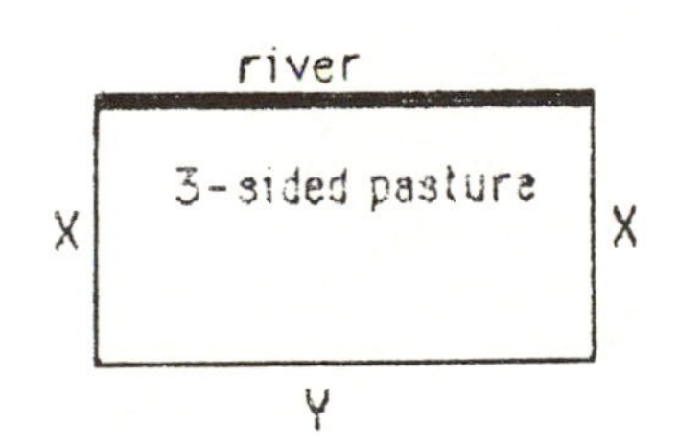

FIG. 14.9.

whether or not that "thing" happens to be real or hypothetical. Data from early child studies show that this assumption is rooted early in perceptual development and hence is likely to function strongly and in a preemptive manner to subvert the appropriate understanding of the representation as a true variable. The reading of pictorial symbol systems simply does not assume that pictures can be representative in nature.

The situation in standard school plane geometry is in this sense parallel, except that in this case one attempts to form hypotheses and prove a result about a class of figures usually on the basis of a single picture, which is often assumed by the student to represent a single referent—even when the student agrees that the figure represents a prototype.

New information technology may change this situation in geometry. For example, *The Geometric Supposer* (Schwartz & Yerushalmy, 1985) provides means by which a student can create a geometric construction on a particular figure that is then stored as a generic procedure repeatable on other figures of that type. The student, in effect, is thus able to vary the picture while holding the construction constant. Moreover, by systematically varying the pictorial representation, one can vary selected features of the pictures, i.e., one can exercise a control of variables in a pictorial context.

We have also observed other strong interactions between the pictorial geometric symbol system and the algebraic symbol system (Kaput, 1982) where we provided a geometry problem to create a set prior to a Students–Professors problem. The performance of students on the translation task who had been given the geometry problem was significantly different from those who had been given an arithmetic problem—indeed, their performance resembled those calculus students described earlier who had been given the pictorial supplement to a Students–Professors type of task.

We expect that study of the differences between geometric and algebraic symbol systems would be especially fruitful, as would the study more generally of the fundamental cognitive processing differences between and interactions among natural and synthetic symbol systems. This is the subject of our current work.

Issues Not Yet Addressed

Despite its length, this chapter has not addressed whole families of issues that an adequate theory of symbol use in mathematics must eventually come to terms with. Such include Pragmatics, the study of the tacit rules that govern symbol use, establish priorities among other rules, determine the scope of certain rules, decide whether or how natural language or other nonmathematical conventions will be applied, and so on. As an interesting thought experiment, consider what would happen to the intelligibility (to a professional logician!) of an introduction to standard predicate calculus if all the usual symbols were replaced by arbitrary ones, if none of the usual conventions held regarding upper and lower case letters, the different conventions distinguishing variables from predicates or the values of those variables, rules of implication from assignment operators, and so on. Assume instead that a different set of conventions was put in effect. At least two things are revealed: the anchoring role of whatever natural language is used and the enormous cognitive importance of the standard conventions, tacit and otherwise.

Another major task is the precise specification that distinguishes in a principled way pictorial–graphical symbolism from alpha–numeric types of symbolism. Along these lines, we have not dealt with the matter of iconicity—see Salomon (1979) for a start on the question. We have not yet given sustained attention to the interactions between media and symbol systems in mathematics, especially regarding computers. The SAM types of systems and *The Geometric Supposer* mentioned earlier suggest that new symbol systems may be possible via the construction of new fields of reference, the vaulting of earlier limitations on the ability of symbols to represent classes or the construction of entirely new kinds of dynamic representation systems. New computer-based representations also support the explicit linking of different representation systems (Kaput, in press).

Finally, we have yet to put theory fully in harness, to plow genuinely new ground or change the conceptual contours of familiar ground. A particularly useful enterprise would be to examine a well-studied domain such as rational numbers and proportional reasoning, which is rich in subconstructs and forms of representation. Ideally, such an examination would help expose in an orderly way how the different subconstructs and procedures fit or conflict with the various symbol systems used to represent them, and most importantly WHY they fit or fail to fit.

To swallow a bit of our own medicine, we could characterize our nascent theory development effort as an attempt to develop a system for representing the myriad of mathematical symbol-use phenomena. Thus its success can eventually be measured in the two dimensions such a representation system has been asserted to function, as a tool of communication, and as a tool to think with.

ACKNOWLEDGMENTS

Preparation of this chapter was supported in part by the National Science Foundation Grant number MDR–8410316. Its contents do not necessarily reflect the views of the Foundation. Many thanks to Ernst von Glasersfeld, Carlos Vasco, and John Richards for thoughtful comments on earlier drafts of this chapter. Its contents do not necessarily reflect their views.

REFERENCES

Barclay, T. (1985). *Guess my rule.* (Computer software) Pleasantville, NY: HRM, Inc.

Baroody, A. J. (1985). Mastery of basic number combinations: Internalization of relationships or facts? *Journal for Research in Mathematics Education, 16,* 83–98.

Bickhard, M. H., & Campbell, R. L. (1985). *A deconstruction of Fodor's anticonstructivism.* Paper delivered to the Annual Meeting of the Jean Piaget Society, Philadelphia.

Bloor, D. (1976). *Knowledge and social imagery.* Boston: Routledge & Kegan Paul.

Bruner, J. (1973). *Beyond the information given.* J. Anglin (Ed.). New York: Norton.

Burton, D. M. (1985). *The history of mathematics: An introduction.* Boston: Allyn & Bacon.

Cajori, F. (1929a). *A history of mathematical notations, Vol. 1: Notations in elementary mathematics.* La Salle, IL: The Open Court Publishing Co.

Cajori, F. (1929b). *A history of mathematical notations, Vol. 2: Notations mainly in higher mathematics.* La Salle, IL: The Open Court Publishing Co.

Carpenter, T. & Moser, J. (1983). The acquisition of addition and subtraction concepts. In R. Lesh & M. Landau (Eds.) *Acquisition of mathematical concepts and processes.* New York: Academic Press.

Carraher, T. N., & Schliemann, A. D. (1985). Computation routines prescribed by schools: Help or hindrance? *Journal for Research in Mathematics Education, 16,* 37–44.

Clement, J. (1982). Algebra word problem solutions: Thought processes underlying a common misconception. *Journal for Research in Mathematics Education, 13,* 16–30.

Clement, J. (1985). Misconceptions in graphing. *Proceedings of the Ninth Conference of the International Group for the Psychology of Mathematics Education,* (369–375) L. Streefland (Ed.) Utrecht, The Netherlands.

Clement, J., Lochhead, J., & Monk, G. (1981). Translation difficulties in learning mathematics. *American Mathematical Monthly, 88,* 286–290.

Davis, P. J., & Hersh, R. (1980). *The mathematical experience.* Boston: Birkhauser.

Davis, R. B. (1984). *Learning mathematics: The cognitive science approach to mathematics education.* Norwood, NJ: Ablex.

Davis, R. B., Jockusch, E., & McKnight, C. (1978). Cognitive processes in learning algebra. *Journal of children's mathematical behavior, 2,* 10–320.

Dickson, W. P. (1985). Thought provoking software: Juxtaposing symbol systems. *Educational Researcher, 14,* 30–38.

Dugdale, S. (1982). Green Globs: A microcomputer application for graphing of equations. *Mathematics Teacher, 75,* 208–14.

Fischbein, E., Deri, M., Nello, M., & Marino, M. (1985). The role of implicit models in solving verbal problems in multiplication and division. *Journal for Research in Mathematics Education, 16,* 3–17.

Fuson, K. & Hall, J. (1983). The acquisition of early number word meanings: A conceptual analysis and review. In H. Ginsburg (Ed.) *The development of mathematical thinking.* New York: Academic Press.

Gardner, H., Howard, V., & Perkins, D. (1974). Symbol systems: A philosophical, psychological, and educational investigation. In D. Olson (Ed.), *Media and symbols: The forms of expression, communication, and education.* Chicago: University of Chicago Press.

Goodman, N. (1976). *Languages of art* (rev. ed.). Amherst: University of Massachusetts Press.

Hawkins, D. (1985). The edge of Platonism. *For the Learning of Mathematics, 5,* 2–6.

Hitch, G. J., Beveridge, M. C., Avons, S. E., & Hickman, A. T. (1983). Effects of reference domain in children's comprehension of coordinate graphs. In D. R. Rogers & J. A. Sloboda (Eds.), *The acquisition of symbolic skills.* New York: Plenum Press.

Janvier, C. (1978). *The interpretation of complex cartesian graphs representing situations: Studies and a teaching experiment.* Doctoral Dissertation, University of Nottingham.

Kaput, J. (1982). *Differential effects of the symbol systems of arithmetic and geometry on the interpretation of algebraic symbols.* Paper presented at the meeting of the American Educational Research Association, New York.

Kaput, J. (1985a). Representation and problem solving: Methodological issues related to modeling. In E. Silver (Ed.), *Teaching and learning mathematical problem solving: Multiple research perspectives.* Hillsdale, NJ: Lawrence Erlbaum Associates.

Kaput, J. (1985b). *Multiplicative word problems and intensive quantities: An integrated software response* (Tech. Rep.). Educational Technology Center, Harvard Graduate School of Education, Cambridge, MA.

Kaput, J. (in press). Information technology and mathematics: Opening new representational windows. *Journal of Mathematical Behavior.*

Kaput, J., & Sims-Knight, J. E. (1983). Errors in translations to algebraic equations: Roots and implications. *Focus on Learning Problems in Mathematics, 5,* 63–78.

Kieren, T., Nelson, D., & Smith, G. (1985). Graphical algorithms in partitioning tasks. *The Journal of Mathematical Behavior, 4,* 25–36.

Kitcher, P. (1983). *The nature of mathematical knowledge.* New York: Oxford University Press.

Kline, M. (1972). *Mathematical thought from ancient to modern times.* New York: Oxford University Press.

Kolers, P. A., & Smythe, W. E. (1979). Images, symbols, and skills. *Canadian Journal of Psychology, 33,* 158–184.

Matz, M. (1980). Towards a computational theory of algebraic competence. *Journal of Mathematical Behavior, 3,* 93–166.

Nesher, P., & Peled, I. (1984, August). *Shifts in reasoning.* Paper given to the Harvard Conference on Thinking, Harvard University.

Nesher, P., & Schwartz, J. (1982). *Early quantification.* Unpublished manuscript. Cambridge: Massachusetts Institute of Technology.

Olson, D. (1985). Computers as tools of the intellect. *Educational Researcher, 14,* 5–8.

Palmer, S. E. (1977). Fundamental aspects of cognitive representation. In E. Rosch & B. B. Lloyd (Eds.), *Cognition and categorization.* Hillsdale, NJ: Lawrence Erlbaum Associates.

Putnam, H. (1975). *Mathematics, matter, and method.* Cambridge, MA: Cambridge University Press.

Reed, H. J., & Lave, J. (1981). Arithmetic as a tool for investigating relations between culture and cognition. In R. W. Casson (Ed.), *Language, culture and cognition: Anthropological perspectives.* New York: Macmillan.

Salomon, G. (1979). *Interaction of media, cognition and learning.* San Francisco: Jossey–Bass.

Schwartz, J. (1976). *Semantic aspects of quantity.* Unpublished manuscript, M.I.T., Cambridge, MA.

Schwartz, J. (1984). *An empirical determination of children's word problem difficulties: Two studies and a prospectus for further research.* Technical Report, Educational Technology Center, Harvard Graduate School of Education, Cambridge, MA.

Schwartz, J., & Kaput, J. (in preparation). *Quantities and mathematics.*

Schwartz, J., & Yerushalmy, M. (1985). The geometric supposer. Pleasantville, NY: Sunburst Communications.

Sims-Knight, J. E., & Kaput, J. (1983). Exploring difficulties in translating between natural language and image based representations and abstract symbols systems of mathematics. In D. R. Rogers & J. A. Sloboda (Eds.), *The acquisition of symbolic skills.* New York: Plenum Press.

Steffe, L., von Glasersfeld, E., Richards, J., & Cobb, P. (1983). *Children's counting types: Philosophy, theory, and application.* New York: Praeger.

Wilder, R. (1981). *Mathematics as a cultural system.* New York: Pergamon Press.

Wollman, W. (1983). Determining the sources of error in a translation from sentence to equation. *Journal for Research in Mathematics Education, 14,* 169–181.

15 The Evolution of Problem Representations in the Presence of Powerful Conceptual Amplifiers

Richard Lesh
WICAT and Northwestern University

Chapter 4 described a process in which problem solving often occurs by: (1) translating from the "given situation" to a mathematical model, (2) transforming the model so that desired results are apparent, and (3) translating the model-based result back into the original problem situation to see if it makes sense and is helpful. Here is an example of this process being used for a standard algebra problem.

> Al has an after-school job. He earns $6 per hour if he works 15 hours per week. If he works more than 15 hours, he gets paid "time and a half" for overtime. How many hours must Al work to earn $135 during one week?

To solve the problem, the preceding "English sentence" description of the problem can be translated into a description using "algebraic sentences."

$$(6 * 15) + 9(x - 15) = 135.$$

Then, a series of algebraic transformations can be used to convert this algebraic model into a form in which arithmetic operations are sufficient to give "an answer." The final transformed description is:

$$x = \frac{135 - (6*15)}{9} + 15$$

This arithmetic sentence can then be solved using a series of arithmetic transformations to reduce to the form: $x = 20$.

So, the entire solution process involves three significant translations: (1) from English sentences to an algebraic sentence, (2) from an algebraic sentence to an arithmetic sentence, and (3) from an arithmetic sentence back into the original problem situation.

Notice that the algebraic sentence that most naturally describes the preceding problem situation does not immediately fit an arithmetic computation procedure. This possibility of "first describing, and then calculating" is one of the key features that makes algebra different from arithmetic.

Here is another example to illustrate the power of translation-transformation "modeling" procedures in algebra.

> A boat, traveling upstream on a river, takes 2 hours to reach its destination 8 miles away. The return trip downstream takes 1 hour and 20 minutes. What is the speed of the river current?

To translate this problem into an algebraic description, let x represent the speed of the boat, and let y represent the speed of the current. Then: $2(x - y) = 8$ and $\frac{4}{3}(x + y) = 8$.

Algebraic transformations *could* be used to solve for x and y in this pair of equations, but another procedure that can be used is to perform a second translation to convert this algebraic model to a geometric model; that is, the graphs of each of the two equations can be plotted. Then, the (x, y) values that satisfy both equations are the same as the coordinates of the point that is on both of the graphs, and the coordinates (5, 1) can be read directly from the graph. So, the algebraic solutions are $x = 5$ and $y = 1$, which means that the speed of the current was 1 mile per hour, and the speed of the boat was 5 miles per hour.

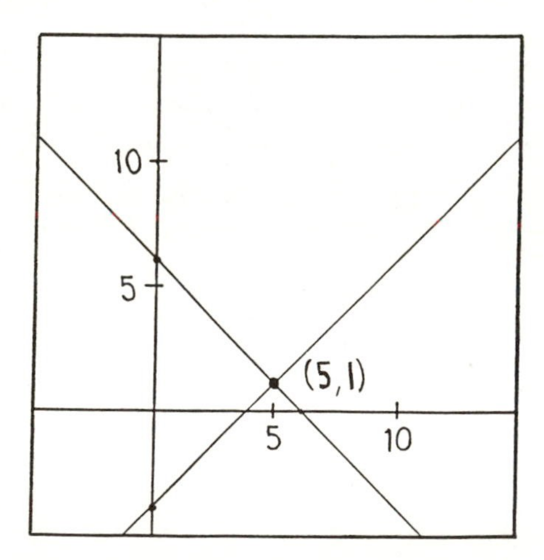

The preceding solution involves three significant translations: (1) describing the problem situation using algebraic sentences, (2) translating the equations into graphic form, and reading the solution from the graph, and (3) interpreting the graphic solution back in the original problem situation.

TRANSPARENT VERSUS OPAQUE REPRESENTATIONS

Most representations can be considered to lie on a continuum between transparent or opaque. A transparent representation would have no meaning of its own, apart from the situation or "thing" that it is modeling at any given moment; furthermore, all of its meaning would be endowed by the student. A relatively opaque representation, on the other hand, would have significant meaning in and of itself, quite apart from that imposed on it by a particular student.

Clearly, many mathematical representations, such as those involving a Cartesian coordinate system, have significant characteristics that are both transparent and opaque. For example, adults today might believe that perceiving the world in terms of a three-dimensional rectangular coordinate system is an intuitively obvious given that is "built into" the world. But the history of science, as well as developmental psychology, proves that many such constructs in fact required a complex system of relationships to be constructed which must be "imposed on" (or "read into") the world before it can seem to be "read out." So, such constructs have important transparent features.

On the other hand, a representation system like the Cartesian coordinate system has power and economy that often actually amplifies the conceptual characteristics of students who learn to use it. Meanings that a student attributes to such "conceptual amplifiers" in one context can serve as "free excess meaning" that is available to help generate new ideas when the same representation is used in another situation. The "boat going up and downstream" problem considered previously can be used to illustrate this point.

Suppose that students have learned to graph linear equations of the form $y = mx + b$, and that a computer will henceforth automatically graph such equations for them, whenever they want. Furthermore, suppose that the students are learning to perform algebraic transformations by giving commands to a computer such as "add 3 to both sides of the equation," or "substitute $x - 4$ for y in the current expression." Then, here is what the algebraic solution to the "boat" problem might look like.

(1)	Divide by 2 to simplify	$2(x - y) = 8$	to	$x - y = 4.$
	Multiply by 3/4 to simplify	$4/3(x + y) = 8$	to	$x + y = 6.$
(2)	Add y and subtract 4 to convert	$x - y = 4$	to	$x - 4 = y.$
	Subtrct x to convert	$x + y = 6$	to	$x = -x + 6.$
(3)	Graph the pair of equations	$x - 4 = y$	and	$y = -x + 6$

(4) Substitute $x - 4$ for y, from the first equation, into the second equation to get

$$x - 4 = x + 6.$$

Then, notice that the two-line graphs from step 3 can be thought of as graphs of the two sides of this equation.

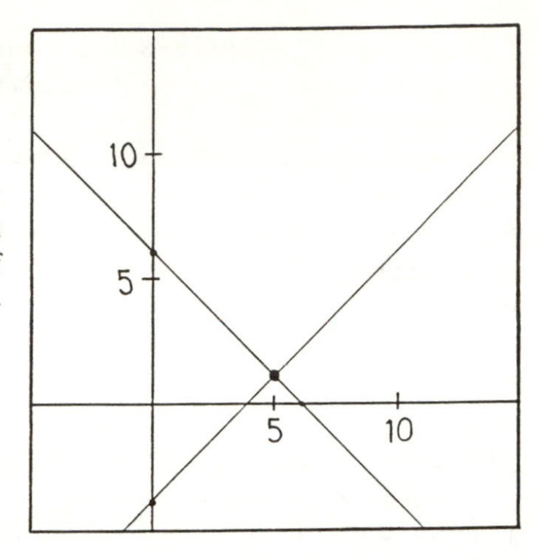

(5) Add x and then add 4 to both sides of the equation to get

$$2x = 10$$

and plot the graphs of the two sides of the equation.

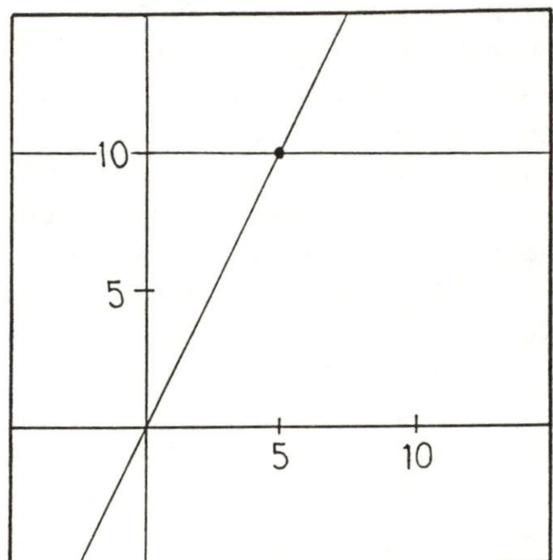

(6) Divide by 2 on both sides to get

$$x = 5$$

and again plot the graphs of the two sides of the equation.

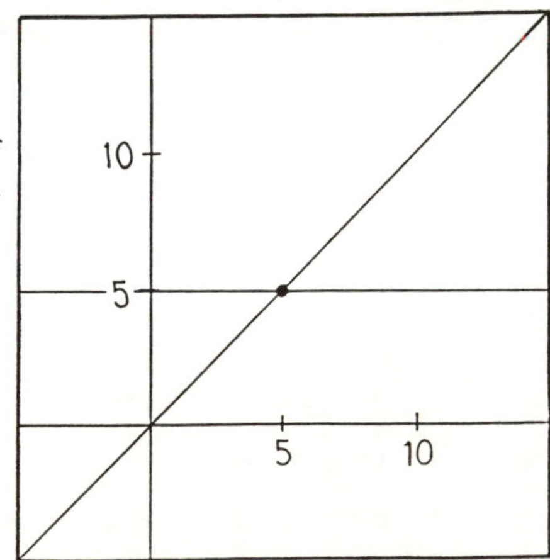

After working on the preceding kinds of activities, many of our students have noticed that, for any given problem, intersection points for the pairs of lines at each solution step always lie on a single vertical line, which turns out to be at the solution point for x. Some of our students have even gone on to think about (and describe) why this invariant feature occurs. So, this dynamic representation system, once constructed, actually helps students to generate significant new questions, and to generate sophisticated solutions related to two of the most fundamental ideas in algebra; that is, our students have used informal language to describe rather deep principles related to: (1) invariance under mappings among isomorphic systems, and (2) invariance under transformations within a given system.

APPLICATIONS AND COMPUTER UTILITIES

Due to the availability of powerful new computer utilities, for the first time, realistic applications can be used to introduce a wide variety of mathematical topics. Rather than first attempting to "teach" a given idea, and then introducing applied problems involving the idea and associated procedures, computer utilities allow us to "give" the procedural capabilities to the student at the beginning of instruction. Realistic applications then can be used to gradually guide students to build their own conceptualizations of the underlying idea (Fey, 1984).

By minimizing the tediousness of answer-giving procedures, computer utilities can focus students' attention on nonanswer-giving phases of problem solving, where activities like data gathering, information filtering, problem formulation, and trial solution evaluation are involved. By focusing attention on the underlying conceptualizations of problem situations, and on the sensibility of products of thought, subtleties about the meanings of relevant ideas become apparent. Also, by reducing the conceptual energies devoted to "first-order thinking," higher order "thinking about thinking" becomes possible. Otherwise students frequently become so embroiled in "doing" a problem that they are unable to think about *what* they are doing, and *why*.

An Aside

Perhaps, before going further, I should state one of my prejudices about the nature of mathematics that is likely to cause confusion for readers who hold different biases; that is, in my opinion, although most mathematical ideas, from early number concepts to algebraic topology, often have computation-type procedures associated with them, "doing the procedure" frequently has little to do with "doing mathematics," nor is it necessarily a good indicator of depth of understanding about the underlying ideas. So, for example, in the study that is described in this chapter, two instructional techniques are compared, one of which minimized the amount of time students spent on "procedure doing" so that attention could be focused on the systems of relationships that constitute the underlying conceptualizations—which, in this case, had to do with simultaneous linear equations.

As an example to illustrate the central point in the preceding paragraph, consider the (mega) ideas underlying differentiation or integration in calculus. The procedures needed to compute a given derivative or integral bear virtually no resemblance to the network of relations that define the underlying ideas and that constitute the meanings. Unfortunately, the "network of relationships" that characterize (i.e., psychologically define) most mathematical ideas are quite difficult to specify. For certain elementary mathematical ideas, considerable progress has been made in this regard in recent years. However, with the exception of certain early number concepts, rational number concepts, and simple spatial/geometric concepts, psychologists and mathematics educators have only begun to map out the specific "conceptual models" that underlie most fundamental mathematical ideas. Certainly very few details can be given about what it means to "understand" a topic like simultaneous linear equations. Consequently, by default, understanding almost exclusively tends to be assessed in terms of students' abilities to carry out procedures—even though "taking a derivative" (say), or solving "one rule under your nose" style word problems, is quite easy to "master" with only the foggiest notion of the meaning of the underlying ideas.

Applications and Problem Solving

Two additional points seem important to mention concerning the nature of applied problem solving, as it is thought about in this chapter: (1) Concerning the goal of making mathematical ideas more meaningful and useful, the results of a recent Applied Mathematical Problem Solving project (Lesh & Akerstrom, 1982; Lesh, Landau, & Hamilton, 1983) support the view that the underlying meanings of mathematical ideas tend to be emphasized during nonanswer-giving phases of problem solving; and using computer utilities, nonanswer-giving phases can begin to be addressed even in conceptual areas characterized by complex number-crunching routines or sophisticated underlying conceptualizations. Consequently, computer utilities can enable us to use realistic applications to introduce whole new categories of sophisticated mathematical concepts; (2) concerning the goal of increasing students' problem-solving capabilities, not only can computer utilities be powerful tools to help students acquire mathematical ideas and increase their meaningfulness, they also can amplify the power of the acquired concepts. With computer-driven "conceptual amplifiers" (like the "symbol-manipulator function-plotter" utility that is described in this chapter, or even familiar tools like VisiCalc), problem solving in the presence of such amplifiers is becoming as important (and perhaps even more so in science, mathematics, and engineering) as that in their absence—although this distinction is artificial. The problem solver no longer can be assumed to be a person working alone with only a pencil and paper for tools. Consequently, assumptions based on such an ''amplified (problem solving) organism'' may have to be considerably different from those common in past cognitive science studies. (New ''realistic'' problem types, such as those involving nondeterministic answers and ''stochastic refinement'' solution processes, also will be increasingly important [Lesh, 1985]).

A Study Comparing Two Computer-Based Instructional Units

The design of the present study was quite simple. Two comparable 10-student groups of ninth graders were given comparable sequences of lessons (six 40-minute sessions) covering the same content, a unit on simultaneous linear equations. Both treatments were computer managed, and both groups saw the same ''slide show'' introduction, which described in general terms what the unit was about, pointing out applications of the ideas to be learned. Both groups also were given the same test, which included items assessing students' abilities to solve pairs of linear equations, as well as some ''applications'' problems similar to those given as examples at the start of this chapter.

The lessons given to both groups included exactly the same examples, exercises, problems, and applications. Differences had to do with: (1) the order in which the various items were given, (2) the part of each activity that was done by the computer (rather than by the students), and (3) the extent to which graphs of solution steps were plotted.

''Utilities'' Group. ''Utilities'' group members were given a ''symbol manipulator/function plotter'' utility that we call SAM. Immediately after seeing the introduction, the ''utilities'' group used SAM's ''magic'' mode to solve three sessions worth of applied problems. In SAM's ''magic mode,'' if a student correctly translated a problem into algebraic sentences, then SAM would step through the equation-solution steps, with each step in the equation-solving process being accompanied by graphs, similar to those in the examples at the start of this chapter. During the first session in the magic mode, the student's task was simply to write a ''correct pair of equations'' to describe each problem. SAM judged the correctness of these descriptions and gave hints if the descriptions were not correct.

During the second and third sessions, SAM did not judge the correctness of problem descriptions; instead, correct or not, the equations were graphed, and (if the student didn't reject the description based on its graph) the solution steps were enacted and graphed one at a time. The students were asked to assess whether the generated solution was sensible or not; and finally, after an assessed answer was judged to be ''sensible,'' SAM verified the correctness of the judgment.

During the fourth lesson, students in the utilities group were taken out of "magic" mode and put into "command" mode. In this mode, SAM would no longer generate solution steps for the students. Instead, students were given a set of primitive commands (e.g., multiply, combine, simplify) and were told that SAM would enact and graph the results of any commands that they would give to SAM. Helps, if students requested them, consisted of SAM going back to the last correct solution step given, and giving one additional step.

The fifth lesson was similar to the first, except SAM operated in "command" mode rather than in the "magic" mode. The sixth lesson was similar to the second and third, with SAM again in the command mode. So, throughout the entire six lesson series, the "utilities" group never had to carry out a single computation step; they simply gave commands telling SAM what to do.

The "Computation" Group. Immediately after seeing the introduction, the "computation" group began learning to compute solutions to pairs of linear equations. After a few examples, the computer gave correct procedural commands and the students were asked to carry out each step, one at a time, until a solution was reached. In a sense, the students in the computation group performed the exact role that SAM had performed for the utilities group; and the computer performed the role that the utilities group students had performed.

Students were checked at each step in the equation-solving process. If errors were made, a hint was given; and if a second error was made on a solution step, the correct response was given and was briefly explained. Graphs were not generated for individual solution steps, although after an algebraic solution was found, the equations were graphed to "check the answer."

The computation group worked through the very same problems as the utilities group, except it took them three sessions to complete the same number of problems as the utilities group had completed in one session.

During the fourth through sixth sessions, the computation group worked as many of the applications problems as they were able. Problem descriptions were checked in a manner similar to the way they were checked in the utilities group; then equation-solution steps were checked one at a time—never allowing the student to take more than one incorrect step without correction.

The computation group students worked individually and completed as many applied problems as they could in the fourth through sixth sessions. In general, they were able to complete less than ¼ as many problems as students in the utilities group—although it of course could be argued that they examined them in much greater detail.

Results

The statistical results can be summarized quite easily. Not surprisingly, the utilities group significantly outperformed the computation group on the applications part of the final examination. More surprising was the fact that they also outperformed their (originally comparable) peers on the *computation* half of the test. Neither group was allowed to use the computer on either half of the final test, and during the instructional sessions, the utilities group had never carried out a single computation step on their own.

The more interesting results came from interviews with the students to tease out any generalizations they had formed that might not have been reflected in any direct way on the test. Some of the most impressive observations that students in the utilities group made had to do with the kinds of "invariance under translations and transformations" facts that were described in the introduction to this chapter. No similar observations were made by anyone in the computation group, nor have I encountered them in years of teaching similar topics to bright college students.

Conclusions

Representations often are assumed to be rather inert entities that only have meanings that are put into them, rather than also having properties of their own. In this study, it was clear that the computer-based representations, which originally had to be constructed and made meaningful by our students, later took on power of their own. They became "conceptual amplifiers" in almost the same sense that the computer itself can be a conceptual amplifier:

1. SAM serves as an expression checker. We didn't have to wait until students gave final answers to problems to know whether they were proceeding along correct solution paths. We could, for example, assess whether they "set up" the equations correctly.
2. SAM is LISP based. It not only generates answers, it produces solution path "traces" that usher in a host of instructional capabilities. It allows us to generate hints by gradually revealing steps one at a time, monitor individual steps in students' solution paths, let students examine processes as well as products of solution attempts, and give students the capability to build/edit/store equation solving routines (like the quadratic formula) in a LOGO-like fashion.
3. SAM's symbol manipulation capabilities interact with its function plotter to produce graphic interpretations of transformations leading to solutions. This gives students ways to visualize symbolic transformations, and (in yet another way) to focus on processes as well as products during solution attempts.
4. SAM can reduce answer-giving phases of problem solving so that attention can be focused on "nonanswer-giving" phases (e.g., problem formulation, trial solution evaluation, the quantification of qualitative information, the examination of alternative possibilities, etc.) where "second-order" (i.e., thinking about thinking) monitoring and assessing functions often are especially important.

Because SAM is not simply an answer-giver, it goes beyond being a tool for *thinking* to become a powerful tool to *think about thinking*. Clearly, these conceptual amplifiers are not simply passive external memories, and equally clearly they will increase the relative importance of nonanswer-giving phases of problem solving like data gathering, information filtering, problem formulation, trial solution evaluation—phases in which "thinking about thinking" or about "the products of thought" are salient.

REFERENCES

Fey, J. (1985). Final Report from the National Science Foundation Award (Report No. SED 80-24425) Washington: The National Science Foundation.

Lesh, R. (1985). Processes, skills, and abilities needed to use mathematics in everyday situations. *Education and Urban Society, 17,* no. 4.

Lesh, R., & Akerstrom, M. (1982). Applied mathematical problem solving: Priorities for mathematics education research. In F. Lester & J. Garofalo (Eds.), *Mathematical problem solving: Issues in research.* Philadelphia: Franklin Institute.

Lesh, R., Landau, M., & Hamilton, E. (1983). Rational number ideas and the role of representational systems. In R. Karplus (Ed.), *Proceedings of the Fourth International Conference for the Psychology of Mathematics Education.* Berkeley, CA: Lawrence Hall of Science.

16 Representing Representing: Notes Following the Conference

John Mason
Open University
England

The main thrust of the conference as I construed it was to seek accommodation between models or descriptions of internal/mental activities, and outer manifestations or representations of those activities; in short, to represent "representing" in some succinct and all-embracing fashion.

Some of my initial confusion with the contributions lay with the word "representation." Kaput's submission gives the standard use of that term in mathematics: A representation requires both a domain to be represented and a codomain to do the representing. (See Kaput for five necessary elements.) From my point of view, it is not clear that "representation" is a sensible or consistent way to describe what goes on inside a person, because their inner experiences *are* their world, and not merely a representation of *the* world, whatever that may be. As von Glasersfeld pointed out, it is more sensible to speak of inner experiences as a person's world, and to speak of their manifestations in terms of pictures, diagrams, words, and symbols as a "presentation" of their world. Furthermore this perspective emphasizes the importance of getting students to use and to become fluent with a variety of modes such as diagrams, symbols, and metaphor to express what they perceive (and not just be confronted with their lecturer's expressions).

At the conference we were all trying to model or describe the inner world of experience. Some of us proceed by contemplating and studying other people, or by studying ourselves as if from outside; others proceed by contemplating and studying ourselves from inside. The exposition of the findings, and in some cases the results, have rather different flavors. I believe that Goldin and Kaput, and to some extent Lesh, proceed from the outside. It therefore makes sense to them to speak of representing how people represent the world. I find it more engaging and realistic to try to capture experience of the inner world from the inside as it were, by using metaphor and descriptive frameworks that resonate with other people's experience. These enable me to observe and modify my own behavior with students, and to speak directly to teachers's experience, so that they too may begin to notice new aspects of their classrooms and consequently begin to modify their own behavior.

The main developments in my own thinking about representation as a result of the meeting come in three areas: First, the nature and role of primitive images (what DiSessa called phenomenological-primitives and I have called core-images) and their place in a world of ideas rather than of materiality; then a clearer picture of the relation between my own sense of students as meaning makers, and a constructivist perspective; lastly, connections between representing and construing.

It so happens that I have recently been seeking fundamental or primitive mathematical images that form the basis of most mathematicians' understanding of specific mathematical topics (at the upper high school/early university level). The more I reflect on DiSessa's phenomenological-primitives, the more I see similarities. There are some close parallels between physics and mathematics, but there are also some distinctive differences as well. Both similarities and differences are worth exploring, and I take this opportunity to make a few steps in this direction. The second area came from von Glasersfeld's superb constructivist articulations, and I have nothing more to add at present apart from its influence on the third area, which emerges as a wish to change the emphasis stemming from the word representing to the emphasis stemming from the word construing. Thus whereas wishing to retain the flavor and direction of Representing Representing as the theme of the conference, I find it more helpful to think of the conference theme now as Construing Construing.

MATERIALITY AND IDEALITY AS PARALLEL REALITY

Natural-science seeks explanatory accounts for observed phenomena. These accounts are invariably in terms of more "basic" or "primitive" ideas, and science educators are now waking up to the fact that we all have "stories" to tell to explain phenomena: My 5-year-old, when asked where electricity comes from, replied immediately "The wall," and when asked how it gets there said "The wall makes it." When asked what it is like, she said "It's long and thin and round." She already has a story to tell (at least when asked), and the story involves reducing the phenomena to more primitive notions such as "something that makes things." DiSessa's chapter gives other instances, and the work of Driver, Easely, Gilbert, and Watts supports this (see Gilbert and Watts, 1983, for a survey). One of the features of more primitive notions used to talk about or explain a phenomenon is that the more primitive the notion, the more idealized it is. Clear examples can be found in DiSessa's contribution (the vacuum-cleaner) and his work on circular motion. In order to give a coherent account of motion, physicists move to friction-less gravity-free world in which particles can have a single force applied to them. Constraints are then added in order to tell a story to explain events in material-world metaphors. It may be the case that untutored, spontaneous, and culturally transmitted primitive notions such as ''resisting a motor makes it work harder'' are more firmly based in the material world, in the form of generalizations of experience rather than simplified ideals. It seems to me that notions such as force and electron are not so much phenomenological primitives as ideational primitives, and this brings the pursuance of natural-science closer to mathematical-science, at least in one respect.

Mathematical-science also takes place in an abstract or ideal world. Points, lines, curves, sets, numbers, functions—none of these are phenomena of the material world, though they do correspond to experiences in the mind. It is possible to say that a coke machine provides an example of a function, that it behaves like a function, but it is not accurate to say that it *is* a function, and the machine is certainly not sufficiently generic to say that it represents or epitomizes functions. Whereas natural-sciences give accounts for phenomena observed in the material world, mathematical-science gives accounts of the structure of the ideal world, and I would go so far as to suggest that it seeks descriptive and manipulable accounts of inner movements/processes (like counting and functioning) that are turned into nouns and then studied as objects that participate in further processes. A formal language is then used to study these ideal objects/processes (in the sense of model theory), and the formal language itself soon becomes identified with the objects of study. This identification goes some of the way to explain the classroom propensity to focus on syntactic rules for manipulating symbols: Most teachers and students are unaware of the existence of anything deeper; they think that the formal language *is* the mathematics. In the early (research phase) work on a class of problems (e.g., rectification of curves, multiplication of numbers, classification of surfaces . . .), there is a struggle to notate and express both the question and its resolution in particular cases. Once one resolution is found, it is abstracted (to investigate the breadth of the class of problems now resolvable), refined (in articulation and in calculation/procedural terms), and simplified (alternative procedures, greater efficiency etc). What emerges is a minilanguage for expressing the problem and its resolution. When this language is taught to students, attention is drawn to the procedural, syntactic aspect, because outer behavior is desired (solving typical problems), visible, and assessable. Thus there is an assessment-induced thrust away from meaning towards mechanical manipulation.

It seems then that both natural and mathematical-science involve a simplified world of ideas—the ideal world. The former uses it to construct stories to account for the material world and the latter gives accounts of the structure of the ideal world. This suggests to me a greater correspondence between DiSessa's phenomenological-primitives and the primitive-images I seek in mathematics than I saw at first, because to be significant science (to contribute to knowledge/scientia rather than to descriptive literature), both must belong to the ideal world. I find this a helpful beginning for further work.

Construing Construing

My reinterpretation of the thrust of the conference intention is to make sense of how we make sense of the world, and to examine the different ways in which different forms of (re)presentation contribute to construing.

Construing, or making sense, or getting-a-sense-of some idea takes place in the midst of manipulating and expressing. By manipulating I mean physically pushing objects such as Cuisenaire rods or Dienes blocks around, or drawing pictures and diagrams in your head or on paper, or metaphorically manipulating abstract symbols. As I pointed out in Mason (1980), manipulation is only possible when you are confident with the "things" as objects. Physical tools can be just as abstractly foreign as marks on paper, and symbols can be as palpable as blocks of wood. Expressing involves trying to articulate, in the most general sense, some inner experience that is quite possibly fuzzy, vague, or ill defined. As I have pointed out elsewhere (Mason & Pimm, 1984), there are big gaps between "seeing" something, being able to "say" something, and being able to "record" that saying on paper in pictures, diagrams, words, and symbols. The importance of struggling to try to say what we See, and then to Record what we Say has been noted so often that it is a cliche to say that only when you try to teach something do you really come to understand it. What is curious is that the cliche has not been incorporated into teaching practice. Very rarely are students actually given enough time to engage in the struggles for themselves. Instead we rush them from initial exposure to written records. Somehow when we are in front of a class, we suddenly believe that if we say "it" to them clearly, they will possess what we have said (present company excepted, of course!).

I found the diagram on the following page a useful presentation of some of the points raised in the various chapters of this volume.

The following are used both as objects to manipulate, and as modes of expression: Physical objects, body movements (muscular and submuscular), pictures, diagrams, images and metaphors (explicit and implicit, pure and material), words and sounds (vocal and subvocal), symbols, formulae.

Lesh uses a diagram to indicate these various forms of representation, with arrows to indicate their use as modes of expression. He claims that teachers have found such a diagram helpful for focusing attention on aspects of the classroom that are generally overlooked. Goldin's diagram is more-or-less dual to Lesh's, as Kaput showed, even though Goldin is describing internal activity while Lesh is describing external behavior. Although I recognize the role and importance of planning and managerial language, I agree with DiSessa that the other distinc-

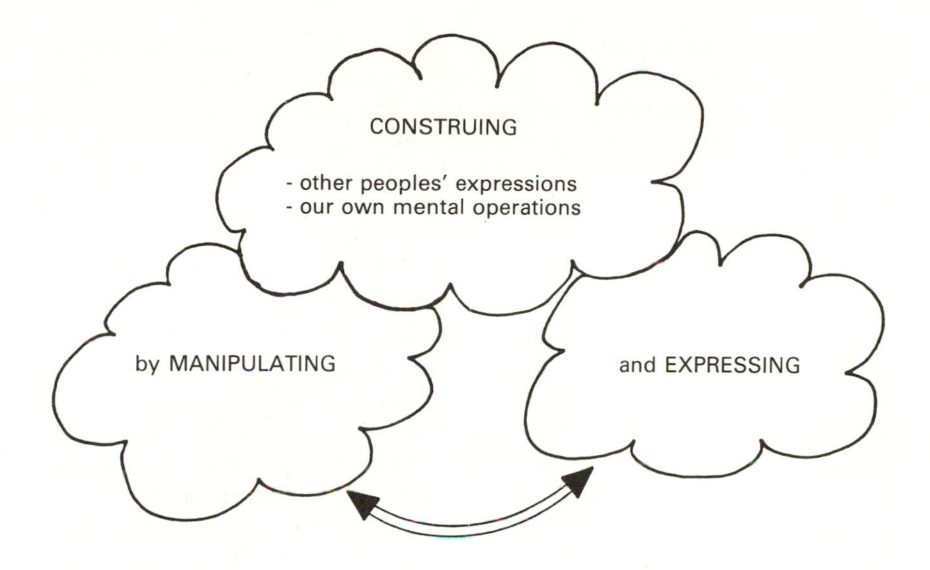

tions are not for me primary. They seem to cut across my experience rather than help me make appropriate distinctions. I may however not be appreciating the distinctions that Goldin is making.

As with any framework, there is the immediate danger that what is intended as an attention focuser will turn into a mechanical scheme for generating yet more useless student activity. That is why I keep stressing the need to ground all such discussion in personal experience, so that teachers find ways of discriminating *their* experience, of noticing and perceiving what happens in class. Teachers can act from such a base. They can modify their behavior in accord with changing awareness. When other agents attempt to institute frameworks in curriculum, the result is yet more routine exercises for students under the guise of fancy terminology.

The to and fro between manipulating and expressing (to one-self or to others) seems to be necessary but not sufficient for constructing meaning, for getting a sense of some idea. In Mason (1984) Pimm and I draw attention to a fundamental question in this area. How can a teacher tell to what a student is attending? For example, the teacher may be ''doing an example'' on the board, seeing through the particular to the general, and expecting the students to attend to that generality so that they can perform the same technique on other examples. All the evidence suggests that ''seeing the general in the particular'' is much more problematic than most teachers realize, largely because teachers are not often aware of what they are doing and so fail to focus attention on the aspects that they are themselves subconsciously stressing, but that students may not be noticing.

There are at least two actions involving construing that deserve attention: Most of the time in class, a student has to try to make sense of what *others* are doing, saying, and writing. There are very specific techniques for studying mathematics from texts (see for example Mason, Burton, & Stacey, 1982)), involving the mathematical processes of specializing and generalizing, and the use of mental images. The same processes are involved in trying to cope with speech and actions. All this is well known and frequently stated, from Plato to the present. Little attention has been devoted to where the basic abstractions of mathematics come from however. Take for example the idea of number. The act of counting engages very young children, as they learn to utter the counting poem peculiar to their parents' natural language. Behind the counting act, as Piaget and Gattegno have observed, lies recognition of matching, of one-to-oneness, and finally, of numbers as objects in their own right. The assignment of word-sounds to numbers is a way of making sense of perceived invariances in the carrying out of operations. No one can abstract the idea of number for a student. Somehow each one of us construes an independent existence of the abstract idea of number, as a result of attending to physical and mental operations that we carry out. Although Piaget has studied these in great detail, and although Gattegno has shown how we can consistently and coherently build on this awareness to develop teaching strategies, most people have concentrated on the external behavior of pupils and focused on trying to classify behavior rather than to develop effective means to aid students to attend to their own inner operations.

The second important action involving construing concerns the items listed under Manipulating and Expressing that are representations, in the sense that they crystallize or give form to inner experiences or conceptions. I find it more helpful to see them as presentations, as manifestations that help to formulate thought by extending the mental screen, rather than as re-presentations of some "thing" already present internally. It is possible, and probably helpful, to train people to be articulate in the various modes. Rather than accept a wishy-washy relativism that salutes the particular strengths of individuals and recommends a shotgun approach in which everyone is subjected to a wide range of images, metaphors, and diagrams, it makes more sense to concentrate on becoming aware of primitive/core images and of effective and ineffective modes of presentation. This is common ground with the suggestion made at the meeting by Kaput, and broadly supported, that we need to study the role and use of all manner of representations so that we know what is effective and what is not. I remain dubious about some aspects of this enterprise as a formal study, because it seems to imbue particular symbols, diagrams, and images with qualities that are for me more properly associated with people. I too argue that there are effective, or poignant, presentations, but I think that I can only learn about effectiveness from the experience of trying to express something to others, and a growing awareness of the very act of expressing. Effectiveness lies partly in the person. I believe that only superficial knowledge will come from statistical studies of "effectiveness" of different forms of diagrams, etc.

My version of the characteristics we discussed for effectiveness runs as follows: (1) It should attract attention to features the expresser wishes to stress and minimize the significance of aspects the expresser wishes to ignore; (2) it should resonate with inner experience of the expresser and with the interpreter.

Having now involved both a person trying to express something and an interpreter who is trying to make sense of the presentation, I am reminded that students spend most of their time trying to construe other people's presentations, rather than learning to present their own ideas. The use of items such as were listed earlier under Manipulating and Expressing are usually taken by students to be faithful representations (in the fullest mathematical sense) of what they need to know. So students concentrate on learning, that is memorizing the presentation. They are probably often blissfully unaware that what they are learning is only partly presented by the diagrams and symbols; that they are being offered windows through which to glimpse or experience something that they too can try to express in a similar mode.

Your feelings, images, etc. *are* your conception of your perception. Thus these various modes of expression are ways of presenting ideas to yourself for reflection and modification. Psychology is plagued by seeking morphisms between the material world and the inner world of experience. This leads to inappropriate questions such as whether mental images exist, and whether (re)presentations are accurate or faithful.

In the Classroom

How can students be helped to appreciate the power and role of symbols, diagrams, etc.? They can be helped to become more articulate in various modes by experiencing the paradigmatic use of powerful symbols such as letters for unknowns and arrows for action. It is not enough simply to use such symbols in the presence of students, however. Explicit attention needs to be drawn to the power of the symbols.

Students *can* be helped to construe more efficiently by having their attention drawn to the basic processes involved. In particular, the processes of specializing in order to find out what an abstract-looking statement says in particular cases, and of generalizing in order to seek the generality being illustrated by particular examples. They can be shown by example that asking yourself what is common among several acts or situations can lead to generality. They can be shown by example that frequently stopping and mentally reconstructing what a text or class has been about helps them construct a reasonably coherent account of a topic, and that by trying to verbalize with the support of diagrams and metaphors they make their ideas accessible on a later occasion. In other words they actually become aware of learning. They can be shown explicitly the role and effect of formalizing (crvstallizing form) in mathematics, which we know as axiomatizing. They can be helped to form mental images, both mathematical and metaphoric, and most especially, to learn how to *dwell in* those images.

The notion of translation from one mode to another carries with it a metaphor of container—you seek what is *in* one expression and put it *in* another format. I don't think you can usefully change your mode of expression without first construing an overall meaning and then reexpressing that in the new mode. Thus translating Caesar into English word by word produces impoverished versions of Caesar's writings. The same will be true of routine exercises in mathematics, such as "translation" from graphical to algebraic or tabular mode. It *is* extremely valuable to associate a kinesthetic and pictorial image with linear, quadratic, . . . functions, to get a feel for their shape and how it is affected by changing parameters. This can be done in a mathematical atmosphere of conjecturing and checking. It is not accomplished by a series of dull exercises.

I noticed that the papers contributed to the conference were centered around three basic metaphors. An architectural metaphor was used in various places to give a sense of building up a picture like building up a house, of constructing meaning from primitive elements. The container metaphor for knowledge is an integral part of the morphism approach to re-presentation, implying that meaning is conveyed from domain to codomain. The voyageur metaphor that I invoked in my submission is another way of depicting the experience of learning and doing mathematics and, in particular, of using the various modes for expressing and giving form to inner experience.

REFERENCES

Gilbert, J., & Watts, M. (1983). Concepts, misconceptions and alternative conceptions: Changing perspectives in science education. *Studies in Science Education, 10,* 61–98.

Mason, J. (1980). *When is a Symbol Symbolic, for the Learning of Mathematics, 1*(2), 8–12.

Mason, J. (1984). *Learning and doing mathematics.* Open University, Milton Keynes.

Mason, J., Burton, L., & Stacey, K. (1982). *Thinking mathematically.* London: Addison Wesley.

Mason, J., & Pimm, D. (1984, August). Generic examples: Seeing the general in the particular. *Educational Studies in Math 15*(3), 277–289.

17 Preliminaries to Any Theory of Representation

Ernst von Glasersfeld
University of Georgia

In contemporary writings on cognition, problem solving, and intellectual development the expressions "representation," "to represent something," and "to represent to oneself" crop up with a certain frequency. People seem to be quite comfortable with them and writers of the most divergent schools of thought use them confidently as though there should be no difficulty at all about their interpretation. Yet, one does not have to look very far in ordinary English texts to find occurrences of the verb "to represent" that show that the word is used with a rather wide range of meanings. Thus, when it is used in technical contexts but without a specific definition, it tends to remain opaque.

Because Goethe was brought into the discussion, I could not help but recall what his Mephistopheles, full of irony, says to the eager student: "Denn eben, wo Begriffe fehlen, Da stellt ein Wort zur rechten Zeit sich ein." (In colloquial translation: "Just where we have no concepts, words come in very handy.")

If, eventually, we want to formulate a *Theory of Representation,* it would seem indispensable that, at the outset, we clarify as best we can what kind of conceptual structure we have in mind when we say "representation."

In the pages that follow I lay out some thoughts in that direction, thoughts that, inevitably, are determined by my conviction that there can be no viable theory of representation without an explicit theory of knowledge. Given the work I have been involved in during the last couple of decades, I am strongly biased in favor of a constructivist epistemology; and these brief notes, therefore, should be taken as an attempt to approach the problem of representation from that particular perspective.

FOUR DISTINCTIONS

My starting point is perhaps best characterized by saying: "A representation does not represent by itself—it needs interpreting and, to be interpreted, it needs an interpreter."

Even a picture is not a picture *of* anything until a viewer (observer, experiencer) relates the colors and shapes he or she perceives in it to the record of things constructed in prior experience. I will interpret a photograph as a picture of myself, *if and only if* I perceive it as similar to images I have at other times perceived when looking into a mirror. But interpretation may also proceed by combining parts. The monsters, for instance, on a canvas by Max Ernst or Hieronymus Bosch do not *represent* "things" in anyone's experiential world. They would be better characterized as *presentations* of originals that these somewhat unusual men cooked up in their imagination, and these presentations are perceived by us as monsters, only because we have experienced lips, eyes, breasts, beaks, claws, and feathers in our own time and are now led to combine these items in novel, unexpected ways. As such, these paintings do not *re*-present but simply provide an occasion to construct something new, and it is up to us, the viewers, to interpret the presentation as creatures of Hell, the Garden of Eden, or caricatures of everyday life.

In discussing representation it seems even easier than in other contexts to produce nonsense. The situation is particularly complicated because the word "representation" is fraught with ambiguity that, for the most part, remains hidden and thus creates untold conceptual confusion. As so often, however, ambiguities surface and become quite obvious when we translate into another language. In the case of representation, I know no better way to lay out the conceptual mess than to escape into German, because there, in different contexts, different words are needed to translate what would be covered by "to represent" in English.

German, in fact, keeps apart the following principal meanings that, in English, are compounded in one word (further distinctions could be made, but for the purpose of this exposition these four suffice):

1. The sketch represents (depicts) a lily = *darstellen*
2. Jane ("mentally") represents something to herself = *vorstellen*
3. Mr. Bush represents (acts or substitutes for) the president = *vertreten*
4. "X" represents (stands for, signifies, denotes) some unknown quantity = *bedeuten*

Given that the German words are usually *not* interchangeable, there are obviously *conceptual* differences that, whether he likes it or not, the German speaker is compelled to keep apart. Because many of the problems we meet in our discussions seem to have something to do with this ambiguity, it would seem helpful to try to separate as neatly as possible the different conceptual structures that become confounded in the fuzzy term *representation*.

The conceptual maze is even more involved than the fourfold division suggests. In sentence (1), "the sketch" (the grammatical subject) refers to the item that *does* the representing. If this grammatical subject is replaced by "the artist," the meaning of "represents" does not change, although it is now not the item called "artist" that does the representing but a sketch or some such implied product of the artist's activity. This type of *agent/activity/product* ambiguity is common in our languages and it is not particularly relevant to the problem of representation. What *is* relevant is that an item such as a sketch, which the linguistic expression purports to be the active agent that does the representing (in the *darstellung/depiction* sense of "representation") is alwavs the result of someone's productive activity.

When we say "This sketch represents a lily," we are expressing a judgment about an experiential item that we have categorized as "a sketch." Our judgment stems from this: Among the constituent operations, which we carry out spontaneously when we perceive (i.e., perceptually construct) this item, there are some that seem so similar to the operations that we spontaneously carry out when we perceive the kind of item we usually categorize as "a lily," that we are prepared to consider the two items equivalent *with respect to these particular perceptual operations*. (Needless to say, such a judgment will always be largely determined by the particular context in which we are operating.)

This, I believe, is compatible with Nelson Goodman's (1976) view. He says: "The making of a picture commonly participates in making what is to be pictured" (p. 32).

Because perceiving, from a constructivist point of view, is always an active making, rather than a passive receiving, the similarity of a picture and what it depicts does not reside in the two objects but in the activities of the experiencer who perceives them. Ordinary language, however, refers to objects as though they *existed* as such, independent of experience. Consequently, it always leads us, the language users, to attribute differences in our perceptual operating to the externalized objects as though they were properties belonging to them in an "objective" sense. Provided we remain aware of this epistemological sleight of hand, we may safely say: An iconic representation (Darstellung) is an artifact and a deliberate reconstruction of another experiential item; the reconstruction selects certain properties considered relevant under the circumstances and uses a medium different from the original.

From the constructivist perspective, the viewer's interpretation of an iconic representation, i.e., what the "icon" will be said to "depict," cannot, as realists tend to believe, be a piece of the "external," ontic world, but only something that, under all circumstances, consists of elements that are already within the viewer's realm of experience. This is the aspect that is dealt with in the next section. In any case, then, such a representation is intended to "stand for" a previously constructed item, but it also differs in some way from that previous construction. It may be simplified or stylized, larger or smaller, two instead of three dimensional, or transformed in some other way.

A *Darstellung* or *icon,* then, has the specific function to "refer" to something else, another experiential item that it is supposed to *depict.*[1]

Lest this be mistaken for a profession of "realism," let me once more emphasize the constructivist approach. A drawing, for instance, will be said to represent a lily, if it is able to produce in the experiencer a reconstruction of the kind of experience he or she has come to call a "lily." But this reconstruction must be somehow different from the construction that yields a "real" lily. If you perceived two lilies on the table, you would hardly consider one of them a representation of the other—in spite of the fact that, in order to be considered "lilies," they must both be constructed by you as the kind of experience that you have come to call "lily." The difference that leads one to distinguish a "real" lily from an iconic representation of a lily may, as I have suggested earlier, be of a variety of kinds, but the one crucial element is probably the realization that there are things that one can do with the lily but not with an iconic representation of one, and vice versa.

A "*trompe-l'oeil*" painting of a lily, or a life-like lily made of plastic, may be intended to trick the beholder into mistaking it for the kind of experiential item that he would spontaneously call "a lily"; if this succeeds, he will say that it *is* a lily, and not merely that it is *like* a lily. In that case, he will assimilate the experiential item to his lily concept without becoming aware of the differences that make it an iconic representation. (Note: if that assimilation is deliberate, it may prompt one specific use of the word "but": e.g., "It's a lily, *but* a plastic one.")

The difference between *darstellen* and *vorstellen* is in part analogous to the difference between a transitive and an intransitive activity. The first begins with a given object, the item that is to be depicted; the second does not begin with an object but creates one. As is so often the case with nouns, *Vorstellung* may refer to the activity or to the activity's result (compare "painting," "diversion," and, indeed, "representation"). In either case, however, *Vorstellung* or "mental representation" refers to a primary creation, to an act of perceptual or imaginal

[1] I use the term *icon* somewhat more loosely than did Peirce in his Theory of Signs (*Collected Papers,* 1931–1935, Vol. 2).

construction, and there is no prior object that serves as "original" to be replicated or *re*-presented.[2]

Hence it would be preferable to move the notion of *Vorstellung* altogether out of the semantic realm of "representation," but, given the present currency of that word, there is no hope that this could be generally accepted. However, in order to keep *mental representations* apart from the others, I accentuate their character of internal construction by referring to them as "conceptions."

In ordinary usage, the things we call "concept" (*Begriff*) often seem to coincide with *Vorstellungen,* but I would prefer to use the term *concept* for those "conceptions" that have been honed by repetition, standardized by interaction, and associated with a specific word. Both are, indeed, retrievable and thus repeatable; but each time one and the same conceptual item is presented it is that item and not a copy or replica of it. In this context it is important to state that, in the constructivist view, "concepts," "mental representations," "memories," "images," etc. must not be thought of as static but always as *dynamic;* that is to say, they are not conceived as postcards that can be retrieved from some file, but rather as relatively self-contained programs or production routines that can be called up and run.

Conceptions, then, are produced internally. They are replayed, shelved, or discarded according to their usefulness and applicability in experiential contexts. The more often they turn out to be viable, the more solid and reliable they seem. But no amount of usefulness or reliability can alter their internal, conceptual origin. They are not replicas of external originals, simply because no cognitive organism can have access to "things-in-themselves" and thus there are no models to be copied.

No matter how new it may seem, a conception is always made up of elements that first arose on the sensory–motor level of experience. Thus, they are made up of elements that the experiencing subject already *has,* though they may, of course, be novel combinations in the same sense that the visual image we construct when we perceive a Max Ernst or Hieronymus Bosch painting is a novel structure in spite of the fact that it consists of well-known parts that we have often used before.

Moreover, these conceptions may, of course, *exemplify* some abstraction—but if they do, they do so by applying the abstraction to quite specific sensory–

---

[2]This is one reason why Kant is so often misunderstood when read in English. In the first sentence of the *Introduction* to his *Critique of Pure Reason,* he rhetorically asks "how could the cognitive faculty be stirred into action, if not by objects which activate our senses and engender *Vorstellungen* . . ."—If, as seems to be the rule, this last term is translated as "representations," the English reader is at once misled into believing that these *Vorstellungen* are to be misunderstood as some sort of pictures of objects. This notion will inevitably be reinforced by subsequent occurrences of representation in spite of the fact that it makes nonsense of Kant's theory of knowledge.

motor material. It is in this sense that Berkeley was right when he said that we could never visualize a "pure generalization" or a "universal" and that, therefore, no such abstraction could "exist." But what Berkeley did not consider was that we could very well retain the way of constructing them, the set of operations that constitute them, or, if you can accept that metaphor, the "program" that produces them. (One's *mental representation* of, say, *one hundred* will be either the numeral "100" or "C," or a specific lot of unitary items whose count is presumed to yield the number word "hundred," or an arrangement of specific lots according to a transform derived from the accepted symbol system, such as "10 × 10".)

In short. a *Vorstellung* or *conception* is a relatively independent conceptual structure in its own right and does not "refer to" or "stand for" something else. But—and to this we return later—it can very well be semantically linked to a word or larger piece of language.

The third use of "to represent" seems to be the least problematic. It is defined in my *Concise Oxford Dictionary* as "Fill the place of, be substitute or deputy for, be entitled to speak for, . . ." By and large it does not create semantic difficulties. It seems to be clear enough in most cases that the item that "represents" another in this way is explicitly designated or empowered to do so on specific occasions and in specifically limited activities.

Even so, it does at times give rise to confusion. When Caligula decided that his horse should "represent" him in the Roman Senate, when a tyrannical governor of Switzerland proclaimed that the people would have to greet his hat as though it were himself, or when nations decreed that a piece of cloth "embodied" their glory and should therefore be saluted, semantic and other conventions got out of hand. This usually happens when a purely "symbolic" representative is turned into an "idol," i.e., a substitute imbued with inordinate power. No doubt such confusions are conceptual, too, but their import is primarily emotional, ideological, or political, and I prefer to disregard them here. With symbols, however, we must deal, but given the German words on which this disentanglement is based, this has to be the topic of the next section.

Bedeuten, in ordinary German, is the word for "to signify" and "to mean" as well as for "to denote." This ambiguity has the same confounding consequences as the ambiguity of "to mean" in English. The German word is used as indiscriminately as the English in sentences such as "These clouds mean rain" and "Hund means 'dog.' " Thus, the difference between an *inductive* experiential relation and an arbitrary *semantic* relation is obscured and the floodgates of erudite obfuscation are opened.

At first glance, one might think that things are not quite as bad with "to represent," because it would sound odd to say "These clouds represent rain." But if we look a little further, it soon gets messy. We come across phrases such as ''His negligence represented a threat to the project'' and ''This letter represents an insult'' where ''to represent'' is used as though it were synonymous with ''to constitute,'' ''can be interpreted as,'' or simply the copula ''to be.'' I could add that these phrases represent the lowest level of semantic precision—and in doing so I would supply yet another current meaning of ''to represent,'' namely ''to be an example of.'' There is no end to the list of possible variations and, although some of them might throw interesting sidelights on the multi-faceted conceptual structure of ''representation,'' I do not intend to pursue their investigation here. That, indeed, is the reason why I put the ''X'' in the fourth of the initial sentences between quotation marks.

I want to consider only the kinds of item that have been deliberately chosen to *represent,* items such as the letters of the alphabet, words, symbols, graphs, and other artificial signs.[3]

There is a widespread confusion about two distinctions that have been made in the categorization of signs, symbols, and other semiotic items. The first is between *iconic* and *noniconic,* and it is analogous to the distinction mentioned in the context of iconic representations in a previous section. The second contrasts ''artificial'' signs with natural ones on the grounds that the former have to be deliberately chosen whereas the latter arise out of the ordinary inductive inferences by means of which a cognitive subject organizes experience. The confusion was generated, at least in part, when the word ''arbitrary,'' introduced to characterize the noniconic items, was slipped into the second distinction as a purported opposite to ''natural'' signs. Whereas it is unquestionably the case that artificial signs are always ''arbitrary'' in the sense that someone deliberately chose them (out of an infinite number of possibilities) to stand for something else, it is equally unquestionable that these artificial signs can be either iconic or noniconic. For instance, the now ubiquitous sign that features a crossed out cigarette is an *artificial* sign irrespective of the fact that it is iconic in that it depicts a cigarette.

Similarly, it was an arbitrary choice that instituted ''X'' as a symbol for ''an unknown quantity'' in mathematical notation. It so happens that it is also *noniconic.*

The coexistence of iconic representations and noniconic or symbolic signs goes back to about 30,000 B.C., the date of the first ''representational'' images of which we know today. They are statuettes of animals and human figures, so obvious that they were at once recognized as ''iconic'' representations when they were found in a German cave in the 1930s. Not so obvious was what Alexander Marshack's recent microscopic examination has shown: These objects were used continuously for many years by their owners, who deliberately carved marks into

[3]For the distinction between ''natural'' and ''artificial'' signs see, for instance, Susanne Langer's (1948) *Philosophy in a New Key* (pp. 58 ff).

them, presumably to record occurrences of some kind of event (Marshack, 1976).

A carefully made scratch, a straight line or angle, was thus chosen to "represent" an instance of a particular experience. Taken individually, there is nothing *iconic* about these marks. Noticing them, perceiving them, gives us not the slightest clue about what they were supposed to "represent." There is no analogy, no correspondence between experiencing the mark and whatever experience they were intended to *refer* to. They are truly *symbolic*, if by "symbolic" we want to indicate that some item was *arbitrarily* chosen to stand for something else.

I want to emphasize that it would be absurd to argue that such a mark should still be considered iconic because, being a single mark, it stands for a single experience. To be considered "a mark," whatever perceivable item one produces must be such that it can easily be isolated from the rest of the perceptual field (if this is not achieved, the item simply fails to function as a "mark"). Hence, the fact that marks, signs, or symbols must be perceived as unitary experiential items intended to refer to a segment of experience that, also, has been isolated from the rest of experience as a discrete and distinguishable piece is simply a prerequisite of marking, signing, symbolizing, and representing. Without it, no semiotic relation whatever could be established.[4]

Thus, one scratch by itself should not be taken as an iconic mark for oneness. The numerical iconicity, however, enters the moment two or more such marks are accumulated to indicate that the marked experience has been repeated so many times. In that case, the sequence of marks iconically represents the number of instances, irrespective of the fact that the marks, as such, give no indication as to what kind of experience was instantiated. Hence, a linear array of three seemingly deliberate scratches on a stone that we believe to be a deliberate iconic representation of a human female is interpreted as someone's record of a repeated experience; but, because the individual scratches are arbitrary symbols, there is no way of deciding whether they individually represent years the female survived, children she gave birth to, dragons slain in her honor, or anything else worth recording in the experience of the statuette's owner.

Thus, already in the beginnings of human culture, we have examples of two kinds of representation, the iconic and the symbolic.

Representations of number do, indeed, provide a complex illustration of what is iconic and what is not. Though the marks on the prehistoric finds and the three first numerals in the Roman system are noniconic with regard to the *kind* of item

[4]Note that this, of course, does not preclude that marks, signs, symbols, etc. can be composites containing any number of elements; it merely means that whatever arrangement of elements is selected and intended to stand for something else must be such that, *in the given perceptual context*, it is likely to be isolated in the perceptual field and taken as a discrete, coherent item.

they are intended to record, they are iconic in the context of numeration, whereas the Roman numerals "V," "X," "L," etc., and the Arabic numerals are not.[5]

Icons *versus* Symbols

Both *refer* to another item, but icons do so by means of sensory–motor similarity whereas arbitrary signs and symbols refer by assignation or social convention. Anyone may infer that a *fleur de lys* is a stylized picture of a lily; no one could infer (by looking at it) that it is the symbol of the Kings of France. This second connection is arbitrary becaue it has nothing whatever to do with the character of specific *perceptual* or sensory–motor operations.

With regard to icons, Piaget's distinction between the "figurative" and the "operative" would seem to be of some importance. Number is not a perceptual but a conceptual construct; thus it is operative and not figurative. Yet, perceptual arrangements can be used to "represent" a number figuratively. Three scratches on a prehistoric figurine, for instance, can be interpreted as a record of *three* events. In that sense they may be said to be "iconic"—but their iconicity is indirect. They do not *depict* "threeness," they merely provide the beholder with an occasion to carry out the conceptual operations that constitute threeness (von Glasersfeld, 1981, 1982). Carrying out these operations does not involve *reference* to some prior sensory–motor item or elements of such items—it is the operating itself that each time constitutes the abstract conception of threeness.

This difference between *figurative icons* that refer to something else and *operative icons* that simply trigger the construction of a specific abstract conception is, I believe crucial in sorting out the kinds of "representation" Jim Kaput (1984) cited from Palmer (1978).

An analogous distinction must be made in the case of symbols. On the one hand, there are symbols that *refer* to figurative items or sensory–motor situations, such as the King of France or the act of smoking; on the other, there are symbols that do not refer to sensory–motor experience at all but are merely indicators that a certain conceptual operation is to be performed. I would call this second category *operative symbols* and would list among them not only number words, numerals, and mathematical signs, such as "+," "−," and "=" but also prepositions, conjunctions, and certain other words whose interpretation does not depend on the recall of sensory–motor experiences but requires the construction of some operative conceptual relation.

---

[5]The fact that Arabic numerals were derived from iconic signs by gradual modification does not make them less "symbolic" today. What matters in this regard is that we and our children do not perceive them as composed of countable items and have to be *told* what they mean.

Final Remarks

A great deal more should be said about the category of *mental representations* I have called conceptions. It is here that one's basic theory of knowledge plays perhaps the most decisive role. From my *radical* constructivist point of view, all the constructs by means of which we assimilate the flow of experience into our "order" or *Weltbild* fall into this category. But I have already exceeded the allotted space and much of what, to me, seems relevant to the discussion of this type of *mental* representation is implicit in other papers I have written.[6]

In any case, these pages should have made it clear that, from the constructivist perspective, it makes no sense to think of mental representations as any kind of *Darstellung* or depiction of ontological reality. From my point of view, the proponents of the various forms of realism throughout the centuries have failed to come up with a viable theory of representation. The recent introduction of the spurious term *information* seems, for the moment, to have revived the old illusion that the gap between conceptual constructs and the ontological world can be bridged. But there is little benefit in speaking of "representations" or, indeed, "translation," where, as Kant's *Critique* has so irrefutably shown, there is no logically possible access to what they are supposed to represent.

ACKNOWLEDGMENTS

I am indebted to John Clement and Jack Lochhead for helpful critical comments about a draft of this chapter.

REFERENCES

Goodman, N. (1976). *Languages of art*. Indianapolis: Hackett.

Kaput, J. (1984, June). *Presentation made at CIRADE Symposium*. Montreal.

Langer, S. K. (1948). *Philosophy in a new key*. New York: Mentor Books. (originally published, 1942)

Marshack, A. (1976). Some implications of the paleolithic symbolic evidence for the origin of language. In S. R. Harnad, H. D. Steklis, & J. Lancaster (Eds.), *Origins and evolution of language and speech* (pp. 289–311). Annals of the New York Academy of Sciences, 280.

Palmer, S. E. (1978). Fundamental aspects of cognitive representation. In E. Rosch & B. B. Lloyd (Eds.), *Cognition and categorization* (pp. 259–303). Hillsdale, NJ: Lawrence Erlbaum Associates.

Peirce, C. S. (1931–1935). *Collected papers* (C. Hartshorne & P. Weiss, Eds.). Cambridge, MA: Harvard University Press.

---

[6]See, for instance, my *Introduction to Radical Constructivism* (1984) and *On the Concept of Interpretation* (1983).

von Glasersfeld, E. (1981). An attentional model for the conceptual construction of units and number. *Journal for Research in Mathematics Education, 12*(2), 83–94.

von Glasersfeld, E. (1982). Subitizing: The role of figural patterns in the development of numerical concepts. *Archives de Psychologie, 50,* 191–218.

von Glasersfeld, E. (1983). On the concept of interpretation. *Poetics, 12,* 207–218.

von Glasersfeld, E. (1984). An introduction to radical constructivism. In P. Watzlawick (Ed.), *The invented reality* (pp. 17–40). New York: Norton. (Originally published in German, 1979)

18 Conclusion

Gérard Vergnaud
Maison des Sciences Humaines

It is not an easy job to comment on the contributions made at a conference for somebody who has not attended it and therefore is not in a position to evaluate the importance of the different points raised in the discussions. Yet it is an interesting challenge, when you know most of the participants and when the topic is so essential.

Representation is a crucial element for a theory of mathematics teaching and learning, not only because the use of symbolic systems is so important in mathematics, the syntax and semantic of which are rich, varied, and universal, but also for two strong epistemological reasons: (1) Mathematics plays an essential part in conceptualizing the real world; (2) mathematics makes a wide use of homomorphisms in which the reduction of structures to one another is essential.

The second point is raised by Kaput when he refers to the use of the word "fundamental" by mathematicians: "most of the results that mathematicians regard as truly fundamental are easily classifiable as representational."

The first point is not so explicit in the papers handled to me. Mathematics does not seem to be viewed so much as a bunch of knowledge concerning space, orientation, quantities and magnitudes, approximations, and as a powerful tool to make predictions about the real world in its physical, technical, economical, and social aspects. And yet this point is essential. We have to face the question "*representation of what, for what purpose?*"

This question is undoubtedly a question for children and students, at all levels, even if it is different for Kaput's college and university students and for Lesh's and Janvier's secondary school students.

The translation problem from one symbolic system to another, from natural language to algebra and back for instance, could not possibly be solved if natural language did not *refer* to the real world and did not convey ideas about properties, relationships, and transformations in the real world. The same is true for algebra, graphs, and other symbolic systems.

The sentence "Peter has won 6 marbles" refers to Peter as a person, to marbles as small round objects to be played with, to the number 6 as the measure of some quantity of marbles, and also to a social game in which transformations (win or loss) may take place.

The graph of "speed against time" refers to different experienced situations and to different ideas concerning movement, time, speed, acceleration, force.

As Douady has put it very clearly in her thesis, mathematical concepts can be viewed both as *tools* to conceptualize new situations and new problems and as *objects* that can be studied in themselves, once they have been introduced. It is easy to understand this point when you think of the introduction of decimals as tools in approximation tasks, and as objects in the analysis of their algebraic and analytical properties. The same is true for negative numbers: In the extension of algebraic manipulations to $a - b$ when b is bigger than a, or in the representation of negative transformations (loss, decrease . . .) and relationships (less than, debt . . .) negative numbers are *tools;* In the theory of abelian groups they are *objects*.

The same is true for higher level concepts like the concept of group itself, or the concepts of function and equation.

Another important point, for me, is the fact that no contributor, except Mason, mentions the distinction between *signified* and *signifier*. Of course one can find paragraphs, in other contributions, in which distinctions of that kind appear, but they are very often confused with others such as the *referent/referred* distinction and the distinction between different symbolic systems (see the *translation* problem from one symbolic system to another one).

I would like to discuss this distinction problem: It has to do with the classifications proposed by Janvier, Goldin, Lesh, and Kaput, and also with the famous distinction made by Bruner between the enactive, the iconic, and the symbolic levels of representation. This distinction underlies the present discussion.

For a cognitive psychologist, whose job is to understand what subjects do and what they say, there is necessarily more than the duality representing/represented. I cannot see any way of dealing with the problem of representation and symbolization without three levels of entities and consequently two problems of correspondence at least.

- the referent } number 1 problem
- the signified } number 2 problem
- the signifier

These words are borrowed from linguists, but they have a different connotation for a psychologist.

The referent is the real world as it appears to the subject along his experience. The world is changing and the subject acts in and upon it to produce events and effects that please him, or that are in accordance with his conscious or unconscious expectations and "representations."

The signified level is at the heart of a theory of representation, in the sense that it is at that level that invariants are recognized, inferences drawn, actions generated, and predictions made. That level is cognitive in essence, as Goldin says: One can also speak in terms of conceptions (Janvier) or in terms of phenomenological primitives (di Sessa). Von Glasersfeld uses the word "presentation" for the German word "vorstellung."

The signifier level consists of different symbolic systems ("schemes" for Kaput, but I find the word misleading) that are differently organized: The syntax of algebra is different from the syntax of graphs, from the syntax of diagrams, and from the syntax of tables. There are even different algebras with different syntaxes, and the graphic syntax for histograms is different from the one for functions. Above all, the syntax of natural language is different, more varied, and more complex than the other symbolic systems, more ambiguous and also more powerful: Natural language is *the symbolic system* par excellence, without which there would probably be no other one.

Of course there are different semantics for different systems and it is essential to recognize the fact that symbols used in communication lie at the signifier level, whereas the meanings lie at the signified level. This means that symbols are usually public (or social), whereas meanings are private (or individual). Very often they are not the same from one individual to the next one, for instance for the teacher and the student.

The correspondance problem between the signifier and the signified is *the problem of the unicity of meaning.* There is also a problem between the signified and the signifier, *the problem of the existence or nonexistence of a symbol expressing a cognitive entity:* It is often the case that one does not have a word or an expression for everything one can think of, nor is it true that all symbolic systems can represent anything. The question raised by Kaput is relevant here: Which aspects of the signified are represented by which aspects of the signified? Graphs are good signifiers for continuity, increase, maximum, or minimum; Formulas are better for calculation.

But all these problems, which stand at the signified/signifier interface, as number 2 problems, would not be solvable, nor even understandable, if there was no number 1 problem at the interface and interaction of the referent and the signified. In other words, the number 1 problem is the problem of adequacy between the signified level of representation (or cognitive level for Goldin) and the real world. Adequacy is never perfect and von Glasersfeld is right in stressing the subjective side of representation: But it is true also that there are parts of the real world that are adequately represented at the signified level as can be proved by the existence of some efficient action in all individuals. The problem of the unicity of meaning and the problem of existence of a symbol to express a cognitive entity have no meaning by themselves. A theory of reference is needed in which the real world and activity in and on the real world are essential.

At this point, I feel the need to make a few comments concerning Bruner's distinction between the enactive, the iconic, and the symbolic levels of representation. Whereas the iconic and the symbolic levels involve signifiers, it is not at all the case that the enactive level involves signifiers: It concerns the signified and the relationship to the referent. The symbolic level can hardly be considered as more difficult that the iconic level, as language (the best example of symbolic system) develops before drawing and reading pictures (the best example of iconic system). The word "enactive" indicates, fruitfully, that representation has to do with action. Let me recall that, for Piaget, representation is primarily interiorization of action, effective action and accommodation, and later, possible action and accommodation.

It is all different, epistemologically, to start from a linguistic approach (signifier, then signified, then referent) or from a psychological approach (referent <-> signified, and then signifier). Research on mathematics education has more to gain from an approach that stresses the essential point: Mathematics is an endeavor to conceptualize the real world in order to act upon it; it is not mainly a language; language and symbols are just a visible part of the conceptual iceberg (see Janvier's metaphor).

A good theory of reference is needed and I am struck by the fact that none of the contributors speaks in terms of situations and actions in situations. When the word "referent" is used, by Kaput for instance, it is mainly in the sense of objects and properties, or in the sense of symbolic systems as organized sets consisting of signs, syntaxes, and semantics. Probably di Sessa's idea of "phenomenolical primitive" is the closest idea to the one I want to stress now.

"Representations (as mixtures of conceptions, know-hows, symbols, and signs) are shaped by situations encountered and mastered by children." Children's and students' behaviors and explanations reveal to us that they have primitive conceptions that take their roots in the first situations they have met and mastered, and in the phenomena then observed. This is true for many conceptions concerning physical, biological, or social phenomena; but it is also true for the ideas of addition and subtraction, multiplication, fraction and ratio, function, transformation, and relationship.

The only difficulty is that not only first situations play a part in shaping representations, but also the whole range of more complex situations that children encounter and progressively master. For addition and subtraction for instance, not only is it the case that the original conception of addition as "something that increases" and subtraction as "something that decreases" have long-term consequences; it is also the case that more advanced ideas including the idea of binary law of combination raise strong obstacles to the understanding of addition of directed numbers. The way from the initial understanding of addition and subtraction of the 4–5-year-old to the full understanding of directed numbers of good 16-year-old students is sided by a large set of different additive and subtraction situations, the mastery of which covers at least 10 years in the development of the student's competences and conceptions. This is true for all mathematical concepts or, better, for all ''*conceptual fields*'' as I name them. The novice/expert distinction is somewhat misleading, because it identifies two states of knowledge in a process where there are many more. It is essentially this reason that has pushed me to propose the theoretical framework of ''conceptual fields'' to study the process of learning mathematics. As a matter of fact, to understand what a student does (or says, or draws) in a situation, it is necessary to know where this situation stands in the whole range of other situations from which this student may draw his ideas.

As a concept refers to more than one kind of situation and as the analysis of a situation requires usually more than one concept, it is a good theoretical and methodological choice to study a set of situations, related to one another in the same conceptual field. Students develop their knowledge inside a fairly wide variety of situations. They catch the simplest properties and relationships first, then more difficult ones, until they master the whole system as a perfectly calculable representation. But before they reach that stage, they do master, and sometimes express, local and noncoherent properties. *Theorems-in-action* is an expression aimed at pointing to the fact that students, in dealing with situations and acting, very often discover and use powerful properties that they are not necessarily able to put into words or into symbols and signs. Nevertheless, theorems-in-action are representational.

A good theory of reference for mathematics education goes together with a good theory of action in situations. This is always the number 1 problem of representation. This does not mean that symbols and signs play no part. As most contributors have mentioned, words, symbols, and signs play a very important part in pointing at the invariants (objects, properties, relationships), and in facilitating the emergence of inferences, rules, and predictions. The approach to symbolization can probably be more profound if the use of symbolic systems is seen within a problem-solving or situation-handling process, as is shown in many examples given by Lesh, Kaput, Mason, and Janvier.

A scheme, for Piaget, is a dynamic totality that ties together all the ingredients of a functional activity and can both accommodate to new situations and assimilate them. The property of universality is already in the scheme, although there is no symbol like x or y to designate the variables. The core of representation at the signified level is probably made of schemes: There are pure sensorimotor schemes like ''climbing a stair case,'' and symbolic sensorimotor schemes like ''enumeration'' (counting a set to find out how many objects there are). Most mathematical schemes concerning space and geometry probably have both kinds of components.

Not only action can be interiorized, but also wording or drawing. And moreover invariants, once fully recognized and expressed, are just as real as a staircase: A function and a vector-space are real objects for a mathematician! This is the main reason why it is impossible to do without a developmental approach to the concept of representation in mathematics education.

The most likely conjecture is that there are at least three types of interactions working at the production of ''representations'' in the student's mind: (1) The referent-signified interaction in which action, chunks, and invariants of different levels, inferences, rules, and predictions play the main part; (2) the signified–signifier interaction in which the natural language and other symbolic systems provide aids for identifying invariants, for reasoning, for planning and controlling action; (3) the interaction between different symbolic systems, as pointed by Janvier, Goldin, Lesh, and Kaput. ''Interaction'' is probably a better word than ''translation'' but interaction requires translation.

It is because symbolic representations may be transparent in different ways (also opaque for some properties of the signified that are not represented), that it is fruitful to use different symbolic systems. Symbolic systems can be ''conceptual amplifiers'' (following Lesh's expression), provided we never forget that they can be misleading, that their use raises specific difficulties, and that they are not the real thing in mathematics.

Author Index

A

B

C

D

Subject Index

D

Q

R